AF347242

# Monsoon Prediction

# Monsoon Prediction

**R. R. Kelkar**
M. Sc., Ph. D.

ISRO Space Chair Professor
Department of Atmospheric and Space Sciences
University of Pune
and
Former Director General of Meteorology
India Meteorological Department

**BS Publications**
A unit of **BSP Books Pvt. Ltd.**

4-4-309/316, Giriraj Lane, Sultan Bazar,
Hyderabad - 500 095
Phone : 040 – 23445605, 23445688

*Published by :*

 **BS Publications**

A unit of **BSP Books Pvt. Ltd.**

4-4-309, Giriraj Lane, Sultan Bazar,
Hyderabad - 500 095
Phone : 040 - 23445605, 23445688
e-mail : info@bspbooks.net

**ISBN : 978-93-52300-70-9 (HB)**

This book is dedicated to
my father

RATNAKAR HARI KELKAR

who had encouraged me to pursue
a career in meteorology

# Preface

Monsoons are observed over many parts of the world, in Asia, Africa, Australia and America, but the Indian southwest monsoon stands out amongst all of them. It is the strongest of all monsoons, it has linkages with the global atmospheric circulation, and it is an important component of the earth's total climate system. The Indian southwest monsoon is India's only source of water. It sustains the livelihood of millions of Indian farmers and influences food production. It is a dominant factor in shaping India's economic growth rate. It has moulded Indian culture and tradition, inspired poets, and set the notes of Indian classical music. The Indian southwest monsoon is indeed "*the* monsoon".

The monsoon is beautiful and captivating, but at times it brings misery and death. It is a regular visitor, but may turn up earlier or later than expected. It makes promises, but does not always keep them. The monsoon rainfall is grossly uneven and India has some of the wettest places on earth and also the driest. The rainfall is not uniform in time either, being interspersed with dry spells. Each year's monsoon is a unique blend of cloud and sunshine and in the strictest sense, it has no past analogues. This is what makes monsoon prediction a scientifically challenging task. The history of monsoon prediction in India dates back to 1886 and it has many records of failure.

Several books have been written about the physics, dynamics and statistics of the monsoon, and some have tried to capture the romance associated with it. Today, with our satellites, models, computers and field experiments, we surely know far more about the monsoon than ever before. The paradox, however, is that our knowledge or appreciation of the monsoon does not necessarily imply our ability to predict it.

This book is not yet another book on the monsoon in general, but it focuses on the problem of prediction. It discusses the current state of art of monsoon prediction, the present and future user requirements, the inherent limitations of science, and why monsoon prediction is a worthwhile scientific effort that needs to be pursued. It covers the different techniques of monsoon prediction on various space and time scales, ranging from mesoscale rainfall to the behaviour of the monsoon across the 21$^{st}$ century. The book is written in such a way that it would be useful to meteorologists and atmospheric scientists and at the same time, not incomprehensible to other readers who may be interested in knowing more about monsoon prediction.

Pune                                                    R. R. Kelkar
August 2008                               kelkar_rr@yahoo.com

# Acknowledgements

In the course of my writing, I have gone through numerous scientific papers, reports and books related to the monsoon and monsoon prediction. Wherever I have quoted facts and figures or discussed important results and methodologies in my book, the authors' names have been mentioned, the sources duly acknowledged and the publication details given in the list of references at the end of each chapter. In particular, I must express my sincere thanks to the India Meteorological Department for the many figures, tables, data and imagery that I have included in this book. IMD has been cited as the source in the appropriate places. I am also very grateful to Dr M. Rajeevan, Director, National Climate Centre, IMD, Pune, for allowing me to quote extensively from his own work and for providing me with resource material.

I would like to place on record my thanks to the publishers of *Current Science* for permission to reproduce two figures from the papers by S. Rupa Kumar et al and R. N. Iyengar et al, published in the journal in 2006. I also thank Dr A. K. Bohra, Head, National Centre for Medium Range Weather Forecasting, Noida, for providing the figures for Section 4.10.

With the intention of ensuring the contemporariness of this book, I have browsed through a lot of information currently available on the internet, but only such figures and data which are in the public domain and without copyright restrictions, have been reproduced in this book. The web site addresses have been provided, so that readers can get their own updates by visiting them directly.

The cover photograph of monsoon clouds at Cherrapunji was taken by my daughter, Ulka Kelkar.

Finally, I wish to express my sincere appreciation to BS Publications, Hyderabad, and particularly to Mr Anil Shah and Mr K. S. Raju, for the timely publication of this book and its very elegant appearance.

R. R. Kelkar

# Contents

(xiii)

# List of Acronyms and Abbreviations

| | |
|---|---|
| AGCM | Atmospheric General Circulation Model |
| AISMR | All-India Summer (or Southwest) Monsoon Rainfall |
| AMEX | Australian Monsoon Experiment |
| AMIP | Atmospheric Model Intercomparison Project |
| AMJ | April-May-June |
| AMMA | African Monsoon Multidisciplinary Analysis |
| AMO | Atlantic Multidecadal Oscillation |
| AMY | Asian Monsoon Year |
| ANN | Artificial Neural Network |
| AOD | Aerosol Optical Depth |
| AOGCM | Atmosphere-Ocean General Circulation Model |
| AR4 | Fourth Assessment Report of IPCC |
| ARIMA | Auto-Regressive Integrated Moving Average |
| ARMEX | Arabian Sea Monsoon Experiment |
| ARPS | Advanced Regional Prediction System |
| BMRC | Bureau of Meteorology Research Centre |
| BOBMEX | Bay of Bengal Monsoon Experiment |
| CC | Correlation Coefficient |
| CCD | Charge Coupled Device |
| CDAC | Centre for Development of Advanced Computing |
| CGCM | Coupled General Circulation Model |
| CLIVAR | Climate Variability Programme |
| COLA | Center for Ocean Land Atmosphere |
| CPC | Climate Prediction Center |
| CTCZ | Continental Trough Convergence Zone Experiment |
| DAC | Department of Agriculture and Cooperation |
| DAI | Drought Area Index |
| DJF | December-January-February |
| DMI | Dipole Mode Index |
| DMSP | Defense Meteorological Satellite Program |
| DST | Department of Science and Technology |
| EASE-Grid | Northern Hemisphere Equal Area SSM/I Earth Grid |
| ECMWF | European Centre for Medium Range Weather Forecasts |
| EMIC | Earth System Model of Intermediate Complexity |
| EMR | Ensemble Multiple Linear Regression |
| ENSO | El Nino Southern Oscillation |
| ENSOW | ENSO Warming Event |
| EOF | Empirical Orthogonal Function |
| EQUINOO | Equatorial Indian Ocean Oscillation |
| FAI | Flood Area Index |
| FGGE | First GARP Global Experiment |

| | |
|---|---|
| FSU | Florida State University |
| GAME | GEWEX Asian Monsoon Experiment |
| GARP | Global Atmospheric Research Programme |
| GCM | General Circulation Model |
| GDP | Gross Domestic Product |
| GEWEX | Global Energy and Water Cycle Experiment |
| GFDL | Geophysical Fluid Dynamics Laboratory |
| GFS | Global Forecast System |
| GPCP | Global Precipitation Climatology Project |
| GPM | Global Precipitation Mission |
| GPS | Global Positioning System |
| ICRP | Indian Climate Research Programme |
| IIOE | International Indian Ocean Experiment |
| IITM | Indian Institute of Tropical Meteorology |
| IMD | India Meteorological Department |
| IMF | Intrinsic Mode Function |
| IMS | Indian Meteorological Society |
| INDOEX | Indian Ocean Experiment |
| INSAT | Indian National Satellite |
| IOD | Indian Ocean Dipole |
| IPCC | Intergovernmental Panel on Climate Change |
| IRI | International Research Institute for Climate Prediction |
| IRS | Indian Remote Sensing Satellite |
| ISMEX | Indo-Soviet Monsoon Experiment |
| ISO | Intra-Seasonal Oscillation |
| ISRO | Indian Space Research Organization |
| ITCZ | Inter-Tropical Convergence Zone |
| JASMINE | Joint Air-Sea Monsoon INteraction Experiment |
| JFM | January-February-March |
| JJAS | June-July-August-September |
| LAM | Limited Area Model |
| LDA | Linear Discriminant Analysis |
| LLJ | Low Level Jet |
| LPA | Long Period Average |
| MAC | Monsoon Annual Cycle |
| MADRAS | Microwave Analysis and Detection of Rain and Atmospheric Structures |
| MAHASRI | Monsoon Asian Hydro-Atmosphere Scientific Research and Prediction Initiative |
| MAM | March-April-May |
| MCZ | Maximum Cloud Zone |
| MESA | Monsoon Experiment South America |
| MJO | Madden-Julian Oscillation |
| MM5 | Mesoscale Model 5 |

| | |
|---|---|
| MMD | Multi-Model Data sets |
| MONEX | Monsoon Experiment |
| MRI | Meteorological Research Institute |
| MSMR | Multichannel Scanning Microwave Radiometer |
| NAM | Northern Annular Mode |
| NAME | North American Monsoon Experiment |
| NAO | North Atlantic Oscillation |
| NASA | National Aeronautics and Space Administration |
| NCAR | National Center for Atmospheric Research |
| NCDC | National Climate Data Center |
| NCEP | National Centers for Environmental Prediction |
| NCMRWF | National Centre for Medium Range Weather Forecasting |
| NHST | Northern Hemisphere Surface Temperature |
| NLM | Northern Limit of Monsoon |
| NN | Neural Network |
| NOAA | National Oceanic and Atmospheric Administration |
| NPI | North Pacific Index |
| OGCM | Ocean General Circulation Model |
| OLR | Outgoing Longwave Radiation |
| PCA | Principal Component Analysis |
| PDO | Pacific Decadal Oscillation |
| PNA | Pacific-North American pattern |
| PPR | Projection Pursuit Regression |
| PRECIS | Providing Regional Climates for Impacts Studies |
| PSLV | Polar Satellite Launch Vehicle |
| QBO | Quasi-Biennial Oscillation |
| RAMS | Regional Atmospheric Modeling System |
| RBI | Reserve Bank of India |
| RMSE | Root Mean Square Error |
| RSMC | Regional Specialized Meteorological Centre |
| SACZ | South Atlantic Convergence Zone |
| SALLJEX | South American Low Level Jet Experiment |
| SAM | Southern Annular Mode |
| SAPHIR | Sounder for Atmospheric Profiling of Humidity in the Inter-tropics by Radiometry |
| SCM | Simple Climate Model |
| SHET | Southern Hemisphere Equatorial Trough |
| SMMR | Scanning Multichannel Microwave Radiometer |
| SOI | Southern Oscillation Index |
| SPCZ | South Pacific Convergence Zone |
| SPIM | Seasonal Prediction of Indian Monsoon |
| SRES | Special Report on Emissions Scenarios |
| SSM/I | Special Sensor Microwave/Imager |
| SST | Sea Surface Temperature |

| | |
|---|---|
| SWAMP | South-West Area Monsoon Project |
| TAR | Third Assessment Report |
| TEJ | Tropical Easterly Jet |
| TIROS | Television and Infra-Red Operational Satellite |
| TMI | TRMM Microwave Imager |
| TOVS | TIROS Operational Vertical Sounder |
| TRMM | Tropical Rainfall Measurement Mission |
| TT | Tropospheric Temperature |
| UKMO | U. K. Met Office |
| VAMOS | Variability of the American Monsoon System |
| VHRR | Very High Resolution Radiometer |
| WCRP | World Climate Research Programme |
| WMO | World Meteorological Organization |
| WRF | Weather Research and Forecasting Model |

# Chapter 1

# The Indian Southwest Monsoon

A Greek traveller named Hippalus is said to have made the discovery of the course of the monsoon winds almost two millennia ago, in the year 45 AD. This led to the opening up of a safe and swift trade route between India and the countries to the west across the Arabian Sea. Traders from countries like Egypt, Syria, Greece, Rome and Arabia would come to Kerala, sailing along with the southwesterly monsoon winds, stay there to buy spices like cinnamon and pepper and other precious commodities, and then return home when the winds became favourable with the retreating monsoon. The existence of a flourishing trade is evident from the Syrian and Jewish copper plates dating back to 850-1000 AD and Roman coins of even earlier years that have been found in archeological excavations in Kerala (Vasisht 2003).

Within India, it is not the monsoon winds but the rains that they bring along with them, that assume a far greater importance. The monsoon is India's only source of water, available for just four months in a year, and required to be stored for consumption over the remaining eight dry months. Farmers have learnt to time their crop seasons in order to get the best out of the monsoon rains. The monsoon has influenced the housing patterns, clothing and eating habits, and lifestyles of millions of Indians. Musical ragas, dance forms and festivals have evolved so as to match the moods of the monsoon. The beauty of the monsoon has inspired poets like Kalidasa and Rabindranath Tagore. The monsoon is literally the central point around which life in India keeps revolving.

From a meteorological point of view, the monsoon is essentially an annual oscillation of the state of the atmosphere in response to the relative position of the sun, as it moves between the tropic of Cancer in the northern hemispheric summer and the tropic of Capricorn in the southern hemispheric summer. However, the monsoon is still looked upon as an extremely complex phenomenon that involves not only the atmosphere, but land and ocean as well. It has attracted the curiosity of scientists from around the world, and of course India, but their understanding of the monsoon is yet far from complete, and the phenomenon is such that it has eluded even a precise and unique definition.

## 1.1 Defining the Monsoon

Sir Edmund Halley was a British mathematician and astronomer who had predicted in the year 1705 by applying Newton's laws of gravitation, that a certain comet that had been seen in 1682 would return in 1758. The comet did indeed return as predicted by Halley and it was later named in his honour. We now know it as Halley's comet and it made its last appearance in 1986.

Edmund Halley is also remembered for his equally remarkable work in a field totally unconnected to comets. Halley had been interacting with mariners and navigators who were acquainted with different parts of India and had lived for a considerable time in the tropics. He had made an extensive analysis of the global patterns of trade winds that blew over the Atlantic, Pacific and Indian Oceans and the seasonal change of wind direction associated with the monsoons. In 1686, he presented to the Royal Society in London, his own holistic view of the trade winds and monsoons (Halley 1686, Figure 1.1.1). He put forth the hypothesis that the monsoon was caused by the differential heating between the Asian landmass and the Indian Ocean. In other words, the monsoon has the character of a giant land-sea breeze that reverses its direction twice during a year. In April, when the sun starts heating the land, the southwest monsoon begins and blows until October; then the land cools and the northeast monsoon blows in the winter until April again (Figure 1.1.2). This was the first ever scientific explanation of the Indian monsoon.

The beauty of Halley's empirical proposition lies in its simplicity. That is why it has survived for more than three centuries and is still talked about. It is difficult to discard it altogether and it is even more difficult to offer an equally elegant alternative. In fact, land temperatures over the Eurasian continent and sea surface temperatures over the Indian Ocean are the two factors that have continued to dominate all efforts to understand and predict the monsoon, but of course in an increasingly complex manner.

To be fair to Halley, we must bear in mind that when he wrote his treatise on the trade winds and monsoons, the present system of global meteorological observations did not exist. Again, his prime intention was to explain only the cause of the large scale seasonal alternation of the trade winds and not the characteristics of monsoon rainfall over India. The need to build upon Halley's basic theory arises from the fact that it is not the process of wind reversal but its byproduct, the monsoon rainfall, which is important to India in practical terms. Any definition of the monsoon must have rainfall factored into it, if it is to be meaningful to the Indian population.

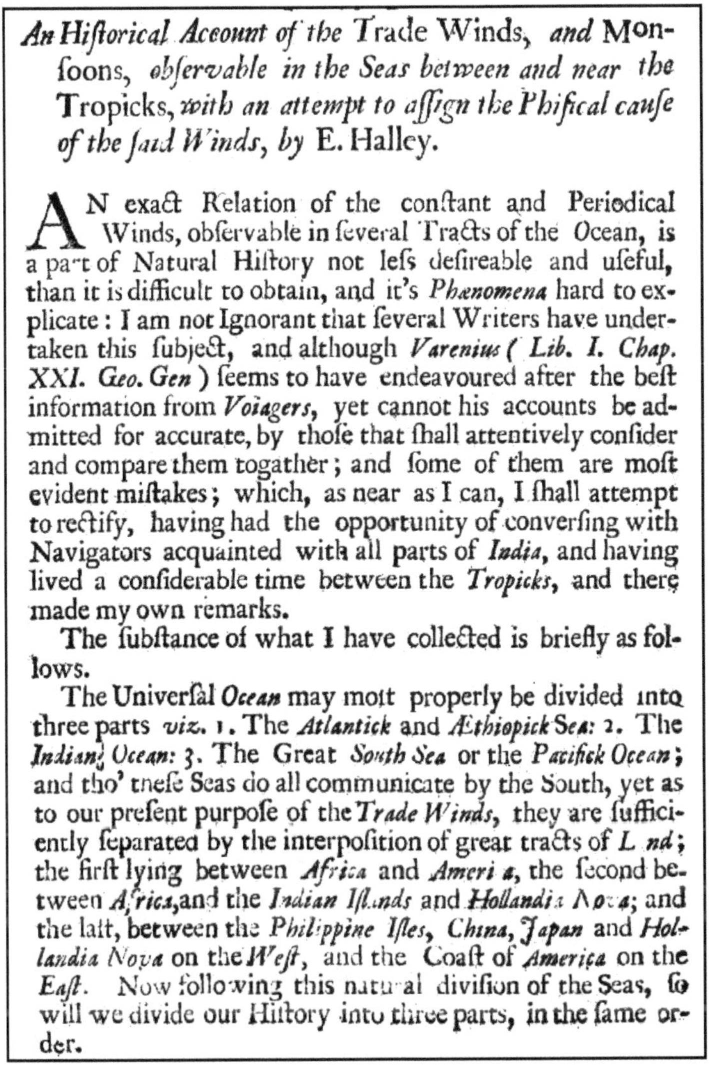

*An Historical Account of the* Trade Winds, *and* Monfoons, *obfervable in the Seas between and near the* Tropicks, *with an attempt to affign the Phifical caufe of the faid Winds, by* E. Halley.

AN exact Relation of the conftant and Periodical Winds, obfervable in feveral Tracts of the Ocean, is a part of Natural Hiftory not lefs defireable and ufeful, than it is difficult to obtain, and it's *Phænomena* hard to explicate : I am not Ignorant that feveral Writers have undertaken this fubject, and although *Varenius ( Lib. I. Chap. XXI. Geo. Gen )* feems to have endeavoured after the beft information from *Voïagers,* yet cannot his accounts be admitted for accurate, by thofe that fhall attentively confider and compare them togather; and fome of them are moft evident miftakes; which, as near as I can, I fhall attempt to rectify, having had the opportunity of converfing with Navigators acquainted with all parts of *India,* and having lived a confiderable time between the *Tropicks,* and there made my own remarks.

The fubftance of what I have collected is briefly as follows.

The Univerfal *Ocean* may moft properly be divided into three parts *viz.* 1. The *Atlantick* and *Æthiopick Sea:* 2. The *Indian Ocean:* 3. The Great *South Sea* or the *Pacifick Ocean;* and tho' thefe Seas do all communicate by the South, yet as to our prefent purpofe of the *Trade Winds,* they are fufficiently feparated by the interpofition of great tracts of *L nd;* the firft lying between *Africa* and *Ameri a,* the fecond between *Africa,* and the *Indian Iflands* and *Hollandia Nova;* and the laft, between the *Philippine Ifles, China, Japan* and *Hollandia Nova* on the *Weft,* and the Coaft of *America* on the *Eaft.* Now following this natural divifion of the Seas, fo will we divide our Hiftory into three parts, in the fame order.

Figure 1.1.1 The first page of Sir Edmund Halley's classical paper on the monsoon published in 1686 (Source: http://www.jstor.org)

The most popularly voiced criticism of the land-sea breeze theory of the monsoon is that while it may be able to explain the genesis of the monsoon, it cannot account for its sustenance, as the temperature contrast vanishes after the monsoon has set in. By implication, the intensity of the monsoon, in particular the associated rainfall, has to be directly related to the land-ocean temperature contrast, but this expectation is also not borne out in reality. On the contrary, it is the land temperatures which appear to get modulated by rainfall, northwest India being the hottest as well as the driest region of the country, both prior to and during the monsoon season (Gadgil 2007). An added difficulty arises from the fact that while monsoon clouds extend up to 10-15 km height, such deep convection cannot be explained by the classical theory (Srinivasan and Joshi 2007). Another inadequacy of Halley's theory that is mentioned often is that it was region-specific. As global monitoring of cloud and rainfall improved over time, it became evident that rainfall over the entire tropical belt and the adjoining subtropics showed an annual oscillation.

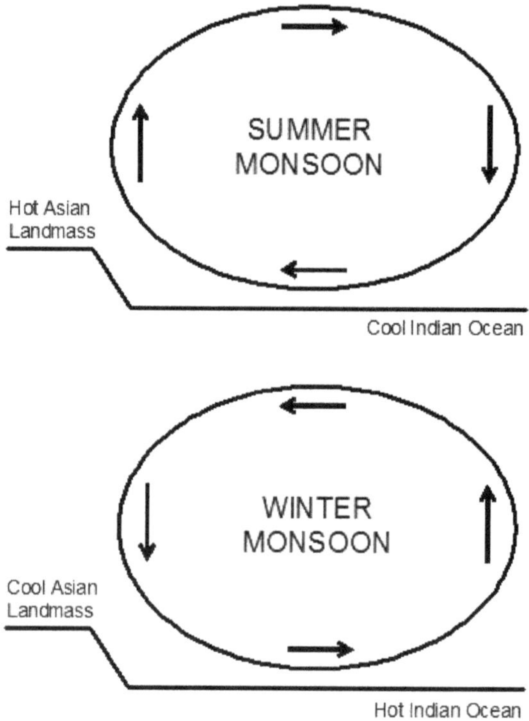


Figure 1.1.2 Halley's concept of the Indian southwest monsoon as a giant land-sea breeze between the Asian landmass and the Indian Ocean

Ramage (1971) made the first attempt to prescribe objective criteria for defining the monsoon and also for identifying the monsoon regions of the world. He postulated that a geographical region could qualify for being called as a monsoon region only if the wind patterns in January and July were distinctly different and satisfied certain basic criteria.

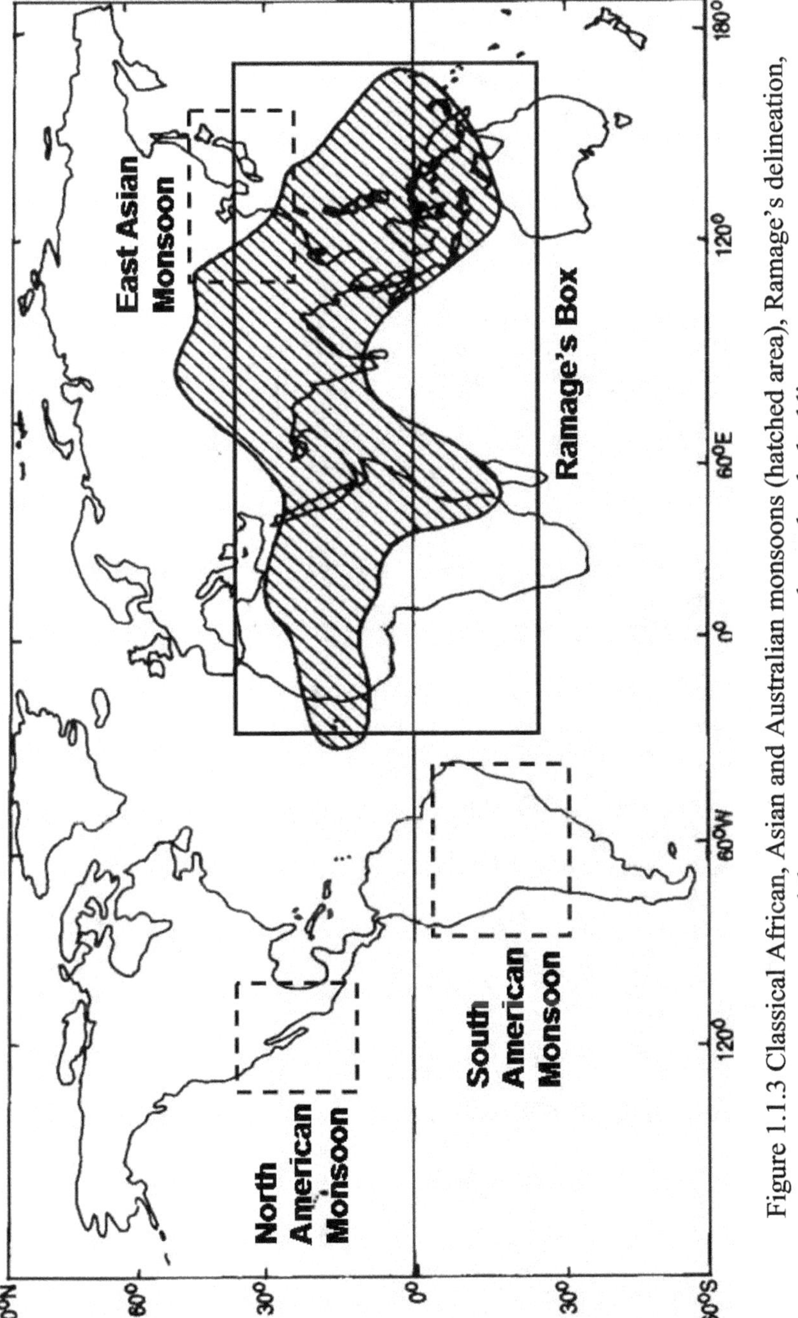


Figure 1.1.3 Classical African, Asian and Australian monsoons (hatched area), Ramage's delineation, and the new monsoon zones shown by dashed lines.

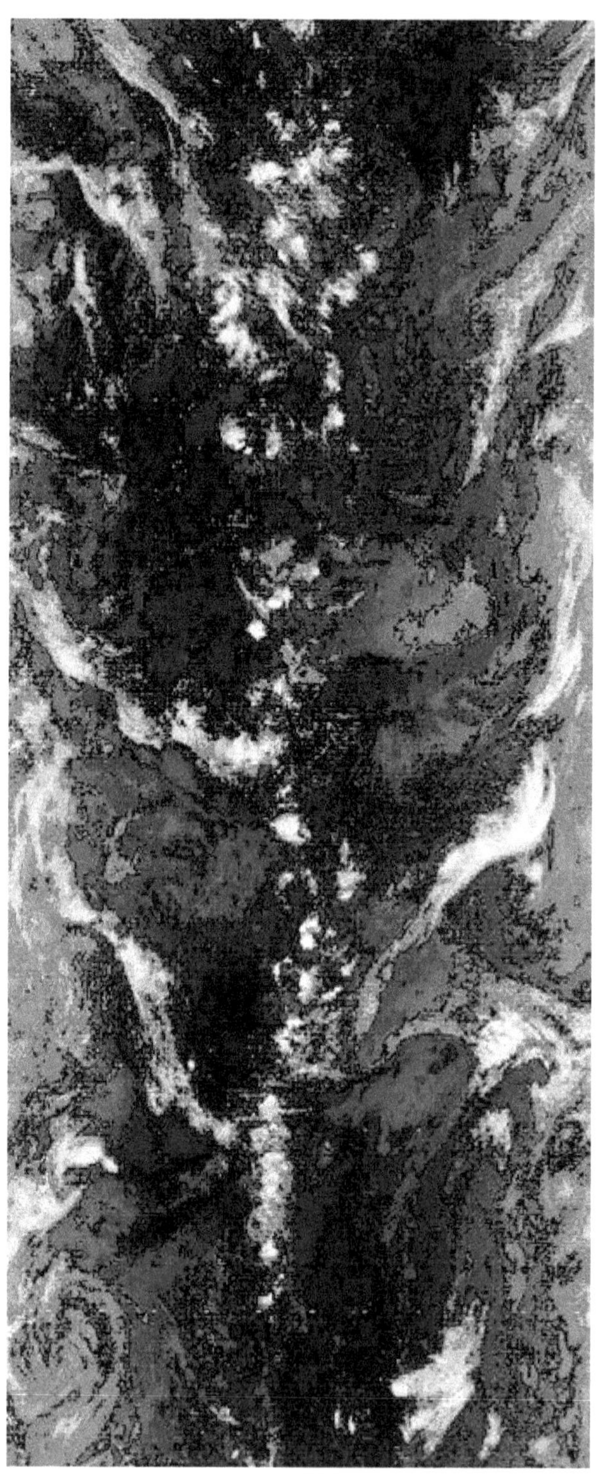

Figure 1.1.4 A typical montage of satellite images showing the ITCZ as a band of cloud clusters encircling the globe.

Ramage's criteria were that the prevailing wind in these two months should blow in a preferred direction at least 40% of the time, they should have a minimum strength of 3 m/sec and there should be a change of at least 120° in the prevailing wind direction between the two months. On this basis, he delineated the monsoon region as the geographical area bounded by the latitudes 35° N and 25° S, and the longitudes 30° W and 170° E (Figure 1.1.3). Although Ramage's criteria were objective, they were still arbitrary to a certain extent, and they restricted the monsoons to south and southeast Asia, northern Australia and tropical Africa. These came to be regarded as the traditional or classical monsoon domains.

Asnani (1993) gave the most inclusive definition of the monsoon in which the role of the Inter-Tropical Convergence Zone (ITCZ) was brought into consideration. The ITCZ that circles the globe (Figure 1.1.4) is a region of lower tropospheric wind discontinuity with horizontal velocity convergence and net upward motion. The ITCZ migrates north-south in association with the march of the sun, and the tropical rainbelt shifts along with it. These seasonal wind reversals and changes in precipitation patterns are not just confined to the traditional monsoon domains, but they also occur elsewhere in the ITCZ region.

There are large regional variations in the seasonal alignment of the ITCZ. Over the eastern Pacific and Atlantic oceans, it remains to the north of the equator throughout the year. In other regions, it moves from the north of the equator in northern summer to the south of the equator in southern summer. Over land, the ITCZ is located over the warmest regions while over the sea, it is located over the highest sea surface temperature (SST) regions. From west Africa to southeast Asia, there is a discontinuity between the westerlies in the near-equatorial region and the easterly trade winds on either side of the ITCZ. The westerlies are largely the southeast trade winds which have changed direction after crossing the equator. Over the Atlantic and Pacific oceans, the discontinuity is between the northeast and southeast trades of the two hemispheres.

Asnani's definition of the monsoon was simple: 'the monsoon is where the ITCZ is', but it paved the way for what could be called the globalization of the monsoon. The monsoon prevails over the whole area of the global tropics and adjoining subtropics (Asnani 2005a, 2005b). Therefore, our present day perspective of the monsoon is not in terms of the seasonal reversal of winds as such, but more in terms of the amplitude of the seasonal change, which is larger over the monsoon region than elsewhere in the tropics. The cause of the larger amplitude is that over the continental regions the land surface heating is determined by the net radiation, whereas the ocean heating is influenced additionally by the winds that drive the ocean circulation. In the non-monsoonal oceanic regions, the location of the maximum SST does not vary with the seasons (Gadgil 2007).

As far as the south Asian monsoon is concerned, the monsoon oscillation is stronger in the northern hemisphere than in the southern hemisphere, and it is stronger over south and southeast Asia than elsewhere in the northern hemisphere. This can be attributed to the Himalayan mountains and the elevated Tibetan plateau producing diabatic heating over a large area of the middle troposphere, the Indian Ocean to the south providing abundant moisture supply and the strong meridional gradients of temperature (Asnani 2005c).

## 1.2 Major Monsoon Domains

In spite of the fact that over the Pacific and Atlantic oceans the ITCZ does not oscillate between the southern and northern hemispheres, the contemporary view about the monsoon is that of a global scale persistent overturning of the atmosphere which occurs over the entire tropics and subtropics and has an annual cycle (Trenberth et al 2000). Embedded within this global monsoon circulation are the more well-known regional monsoons of Asia, Australia and Africa. Against the concept of a global monsoon, the traditional idea of a large scale thermally driven land-sea breeze appears to be too idealized. The wind reversal factor is hardly evident in some of the newly identified monsoon regions.

## 1.2.1 Asian Monsoon

The Asian monsoon in its totality is the largest monsoon system of the world. Over land, the Asian monsoon region includes the Indian subcontinent and the Indo-China peninsula, extending northeastward into mainland China, Korea and Japan. Over the ocean, it covers the South China Sea and the northwest Pacific Ocean. The Asian summer monsoon system incorporates two large regional subsystems: the Indian southwest monsoon or the south Asian monsoon and the east Asian monsoon. These two subsystems are partly interrelated, but they also exhibit independent behaviour at times. The phenomenon of the Asian monsoon has been treated exhaustively by Asnani (2005d) and a concise description of the east Asian summer monsoon has been given by Kripalani et al (2007).

The most significant feature of the east Asian summer monsoon is the quasi-stationary front that is seen to extend from south China to Japan. This front is known by different names in different countries: as Mei-yu in China, Changma in Korea and Bai-u in Japan. It is located along the northwestern periphery of the north Pacific subtropical high. The alignment and strength of this high influences the monsoon. When the high extends more to the west it intensifies the low level jet to its northwest, bringing more moisture into the Yangtze river basin and increased monsoon precipitation. The west Pacific warm pool which is situated at the southern edge of the high also influences the east Asian monsoon.

The climatological mean date of onset of the monsoon over the South China Sea is around the middle of May. This results as a combination of the tropical deep convective rainfall over the equatorial region and the rainfall associated with the subtropical front over south China. After this onset has taken place, the monsoon rainbelt runs from the Arabian Sea across the Bay of Bengal, through the South China Sea, to the northwest Pacific Ocean. It is associated with the characteristic southwesterly winds that connect the south Asian and east Asian monsoons. The monsoon then advances to the north and northeast across the Yangtze river basin into south Japan by middle of June. It subsequently enters north China, Korea and north Japan. The advance of the east Asian monsoon does not proceed at a steady pace and it could be very rapid or extremely sluggish in phases. The southeastern plains of China receive a mean annual precipitation of over 150 cm. The northwestern regions are much drier, the Gobi desert getting only 25 cm of rain, and in some parts even less than 10 cm, annually.

## 1.2.2 Australian Monsoon

North Australia has a predominantly monsoon climate with 90% of the annual rainfall occurring during the southern hemispheric summer months from November to April. The north Australian monsoon is caused by the temperature contrast between the heated Australian mainland and the cooler waters of the Pacific Ocean to the north of it. A heat low develops over northwest Australia and the ITCZ runs through it. The maximum monsoon rainfall occurs over the region of confluence of the ITCZ and the South Pacific Convergence Zone (SPCZ). The Australian monsoon is triggered by cold surges from the South China Sea and surges in the low level southerlies along the west Australian coast. The deep tropical convection over north Australia and Indonesia provides the upward branch of the Pacific Ocean Walker circulation which governs the southern oscillation (Section 2.2).

There are many commonalities between the Indian southwest monsoon and the north Australian monsoon (Asnani 2005e). The north Australian monsoon has a well-marked onset (Hendon et al 1990) and withdrawal. There are intraseasonal and interannual variations in the intensity of the

Australian monsoon (Holland 1986). There is a monsoon trough, equatorward of which there are lower level westerlies and upper level easterlies, and monsoon depressions form over the adjoining oceans. Although the characteristics of the Australian monsoon quite resemble those of the Indian southwest monsoon, the absence of tall mountains like the Himalayas in Australia makes a lot of difference and the diabatic heat supply from the land to the atmosphere remains confined to the lower levels. The net result is that the Australian monsoon is not as intense as the Indian southwest monsoon. Many specific features of the Australian monsoon were investigated during the BMRC Australian Monsoon Experiment (AMEX), details of which have been described by Holland et al (1986).


### 1.2.3 North and South American Monsoons

There are parts of both north and south America which receive more than half of their annual rainfall during the summer months, with a relatively dry winter season, like in other summer monsoon regions of the world (Asnani 2005f). They also exhibit many of the basic features that characterize other major global monsoon regimes, such as the land-sea temperature contrast, a thermally direct circulation with a continental rising branch and an oceanic sinking branch, land-atmosphere interactions, intense low level inflow of moisture into the continent, and associated seasonal changes in regional precipitation. As most of south America lies within the tropics, however, the precipitation amounts are comparatively far higher, and seasonal temperature differences are much less pronounced than over north America.

The north American monsoon is also popularly known by other names such as the southwest U. S. monsoon, Mexican monsoon and Arizona monsoon. The wet season begins in July and continues up to mid-September. The core of this monsoon is actually situated over northwestern Mexico, but it influences much larger areas of the southwestern U. S. It is driven by the warm land surface areas of this region together with atmospheric moisture supplied by nearby maritime sources such as the Gulf of Mexico and the Gulf of California. The north American monsoon rainfall exhibits considerable variability on different temporal and spatial scales.

Both the north and south American monsoon regimes show a distinct life cycle from onset to decay, the onset usually being characterized by a sharp change from hot, dry conditions to cool, wet ones. The onset phase of the north American monsoon is characterized by the rapid northward spread of heavy rains, the mean dates of onset ranging from early June over southwest Mexico, to early July over Arizona and New Mexico. There is increased vertical transport of moisture by convection and southerly winds flowing up the Gulf of California. The south American monsoon makes its onset over the equatorial Amazon from where it spreads quickly eastward and southeastward across the Amazon basin. The onset of the wet season in Brazil typically occurs towards the end of September or early October. The rapidity of the onset of the south American monsoon suggests a more dynamical origin rather than just a seasonal change. The period of the heaviest monsoon precipitation is July-September over north America and November-February over south America. At this time, a band of deep convection known as the South Atlantic Convergence Zone (SACZ) is seen to extend from central to southeast Brazil on towards the Atlantic Ocean. SACZ is the Atlantic counterpart of the SPCZ which is important to the Australian monsoon.

Compared to the south American monsoon, the north American Monsoon has been investigated much more extensively (Douglas et al 1993, Adams et al 1997). An experiment called the South-West Area Monsoon Project (SWAMP) had been organized in 1990 to observe the convective systems over Arizona and Mexico and study the monsoon environment. Currently a major experiment called the Variability of the American Monsoon System (VAMOS) is in progress under the umbrella of the Climate Variability (CLIVAR) programme of the World Climate Research Programme (WCRP). VAMOS is a concerted attempt towards achieving a better understanding of the north and south American monsoon systems in their totality (Higgins et al 2003, Nogues-Paegle et al 2002, VAMOS 2008).

VAMOS has launched two complementary international programmes called the North American Monsoon Experiment (NAME) and Monsoon Experiment South America (MESA). The common objectives of NAME and MESA are to understand the key components of the two American monsoon systems, their variability and their role in the global water cycle, to build observational data sets and to improve the simulation and monthly to seasonal scale prediction of the monsoons. It also has subexperiments and field campaigns such as the South American Low Level Jet Experiment (SALLJEX) designed to study specific phenomena like the low level jet.

The results of NAME and MESA have already started providing new insights into various aspects of the American monsoon systems such as moisture transport processes, structure and variability of the South American low level jet, and the diurnal cycle of precipitation in the core monsoon regions. They are also helpful in model development and hydrological applications. As research on the American monsoon systems proceeds further, a unified view of the climatic processes modulating continental warm season precipitation is expected to emerge (Vera et al 2006).

### 1.2.4 African Monsoon

An exhaustive description of the monsoon systems that prevail over different parts of Africa has been given by Asnani (2005g). The region of central Africa between 15° N and 15° S receives the highest rainfall over the continent while Sahara in the north and Kalahari and Nahib in the south are desert areas. The coastal strip of north Africa receives its rainfall in association with extratropical westerly waves and low pressure systems which move across the Mediterranean Sea in winter. Over west Africa, the ITCZ has a north-south oscillation, but it remains north of the equator throughout the year. However, the air on both sides of the ITCZ being dry, there is no rain in the vicinity of the ITCZ. In northern hemispheric summer, the ITCZ reaches its most northerly location giving rains over Sahel and areas to its south up to the Gulf of Guinea. When the ITCZ moves southwards, Kenya and Uganda receive what are called 'short rains' in the months of October to December and as it goes northwards once again, they get 'long rains' from March to June. Tanzania gets its rains between November and March. The onset and withdrawal of the summer monsoon do not have a clear association with the migration of the ITCZ because of the influence of major orographic features like Lake Victoria and the Great Rift Valley and local topography.

Sanjeeva Rao and Sikka (2007) have discussed the commonalities and differences, as well as possible interactions between the African and Indian monsoon systems. Both monsoons develop in close proximity and have similar large scale atmospheric features, but the Indian monsoon is much stronger than the African monsoon and extends further north in the peak phase in July. The Indian monsoon is dominated by a 30-50 day oscillation and there is a possibility of its being modulated by an eastward propagating Madden-Julian Oscillation (MJO), when travelling over the near-equatorial African region. This interaction can be detected through eastward moving low outgoing longwave radiation (OLR) pulses and corresponding fluctuations in the 250 hPa velocity potential. A comparison of the occurrence of drought over India and the Sahelian and sub-Sahelian regions of Africa has shown that although the drought years may not exactly match, there is a signal of a common multi-decadal mode in the monsoon variability over the two regions.

An international project known as the African Monsoon Multidisciplinary Analysis (AMMA) is currently under implementation. This major initiative is aimed at improving the knowledge and understanding of the west African monsoon and its variability on time scales ranging from daily to interannual. The AMMA project is motivated by an interest in fundamental scientific issues and also by the need for an improved prediction of the west African monsoon. Since 2006, over 400 scientists from more than 25 countries, representing more than 140 institutions have been involved in AMMA.

The present observational and monitoring system over the west African region is inadequate for the purposes of numerical prediction and also for studying the diurnal, seasonal and annual cycles of monsoon rainfall. From a wider perspective, the latent heat release through deep convection in the ITCZ over Africa is a significant heat source that affects the global circulation, including Atlantic Ocean hurricanes, many of which have their origins in west African systems. AMMA aims at investigating the west African monsoon in a holistic manner, including the role played by the Saharan region which is the world's largest source of atmospheric dust and aerosols. One of the objectives of AMMA is to study the phenomenon of the west African monsoon from the global and regional scales down to the mesoscale and submesoscale and how the scale interaction results in the observed features of the monsoon and its variability. The latest information on this project can be obtained from the AMMA web site on the internet (AMMA 2008).

## 1.3 Indian Southwest Monsoon

Asnani (2005h) describes the Indian southwest monsoon as the most intense among all the monsoons of the world. In summer, the Himalayan mountain range and the Tibetan plateau act as an intense and extensive heat source that supplies heat directly to the middle troposphere. In winter, they act as a heat sink. The resulting annual oscillation in pressure, temperature and wind is unique to this region, and makes the southwest monsoon the strongest monsoon. In addition, under the influence of orography, the southwest monsoon produces some of the world's highest precipitation amounts. Cherrapunji and Mawsynram in India are known to be the rainiest places on earth.

By the 1960s, a fair understanding of the Indian monsoon system had been achieved, but this was primarily based upon the surface and upper air observations made routinely at land stations over the Indian subcontinent. The role played by the ocean in the monsoon processes became clearer subsequently when systematic observations over the sea became available through four international experiments. These were: the International Indian Ocean Experiment (IIOE 1961-1964), the Indo-Soviet Monsoon Experiment (ISMEX 1973), the International Monsoon Experiment (MONSOON 1977) and the Monsoon Experiment (MONEX 1979). MONEX-79 was a subprogramme of the First GARP Global Experiment (FGGE), which in turn was a part of the Global Atmospheric Research Programme (GARP).

MONEX was an international effort that led to the collection of extensive surface and upper air observations over land stations, supplemented by observations made during special ship cruises and aircraft reconnaissance flights. A geostationary satellite had also been specially positioned over the Indian Ocean. MONEX-79 stands out as a landmark effort in monsoon research, and the data set generated by MONEX helped scientists all over the world to undertake studies on the Indian monsoon for many years thereafter. For much of what we know today, such as the monsoon onset processes, the onset vortex, the low level Findlater jet, the Somali current, monsoon inversions over southeast Arabian Sea, air-sea interactions over the Indian Ocean, and so on, we owe a debt of gratitude to MONEX.

The oceans had been a generally data-sparse region for long, but particularly so in the case of rainfall. It was only after the launch of geostationary meteorological satellites like INSAT, that it became possible to indirectly estimate the large scale precipitation over the oceans (Arkin, Rao and Kelkar 1989) and the real breakthrough came with the Tropical Rainfall Measurement Mission (TRMM) satellite that carried an onboard precipitation radar. Under the Global Precipitation Climatology Project (GPCP), rain gauge measurements and satellite-based rainfall retrievals have been skillfully blended and extensive rainfall data sets and statistics have been compiled. With the help of the GPCP global analyses it is now possible to quantify the seasonal alternation of the precipitation patterns in the monsoon domains of the world, something that Halley could only have conceptually visualized in 1686, in the absence of any data to support his argument. The GPCP maps which currently cover the data period 1979-2007, clearly depict how in January (Figure 1.3.1), the Indian subcontinent is almost devoid of any significant precipitation, the rainbelt associated with the ITCZ having shifted to the south of the equator over the Indian Ocean. In July (Figure 1.3.2), on the other hand, some of the rainiest areas of the world are over the Bay of Bengal, northeast India, and the west coast of India in association with the southwest monsoon.

Figure 1.3.3 and 1.3.4 show the mean sea level pressure patterns for January and July respectively. The steep south-north pressure gradient that develops over India during the monsoon season is absent is winter. Figures 1.3.5 and 1.3.6 depict the low level wind flow at 850 hPa. In July the crossequatorial flow and the strong southwesterlies of the monsoon are clearly evident, whereas in January the northeasterlies over the Bay of Bengal and the southern Indian peninsula are seen. A striking feature of Figures 1.3.7 and 1.3.8 is the strong westerly flow at the 200 hPa level in January being replaced by the tropical easterly jet in July in association with the monsoon. (*Henceforth, in this book, the word 'monsoon' will refer to the Indian southwest monsoon unless stated otherwise.*)

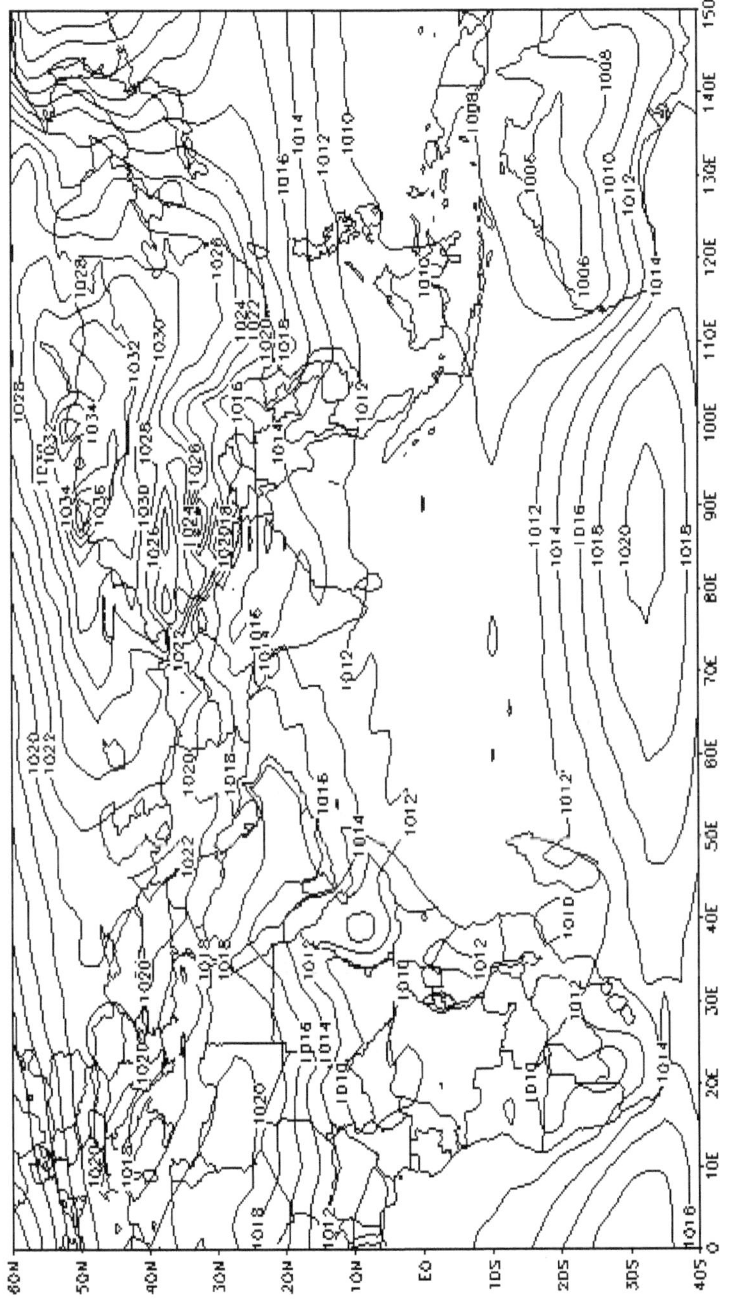

Figure 1.3.3 Mean sea level pressure (hPa) in January averaged over the period 1979-1995
(Source: IMD)

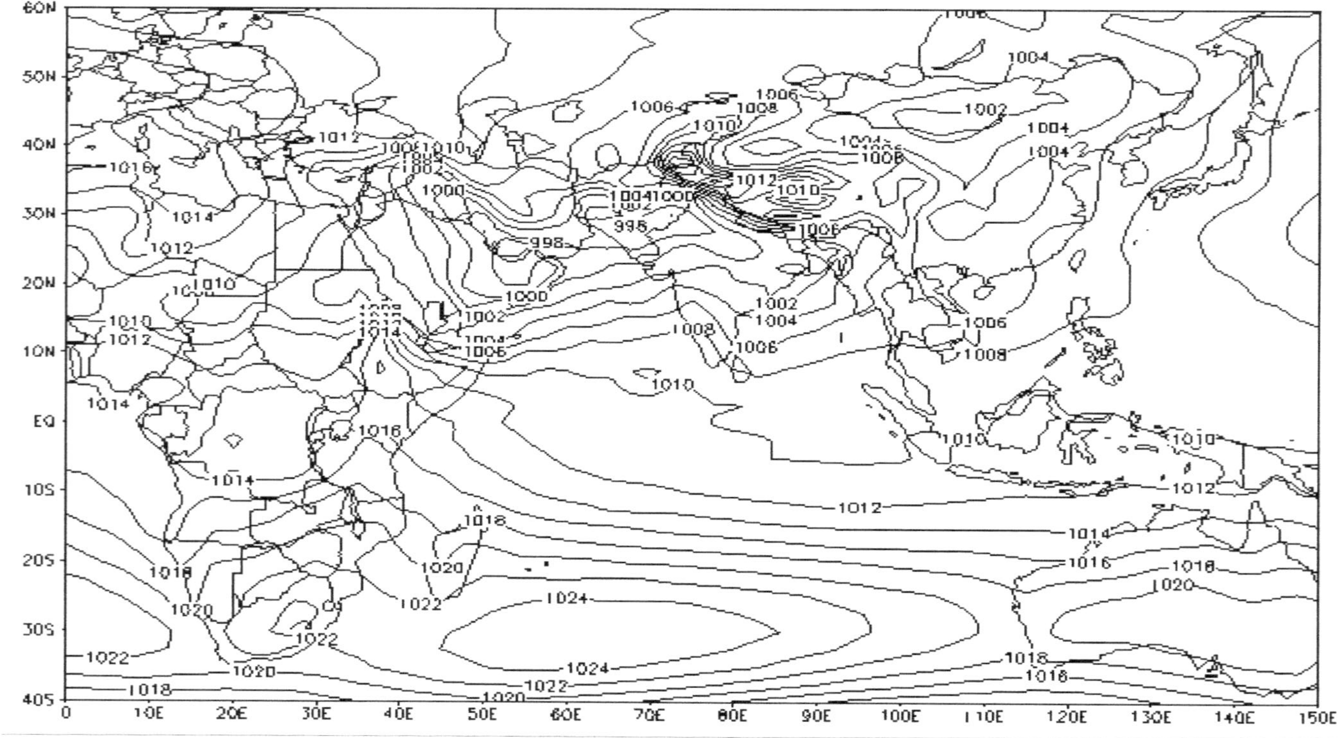

Figure 1.3.4 Mean sea level pressure (hPa) in July averaged over the period 1979-1995
(Source: IMD)

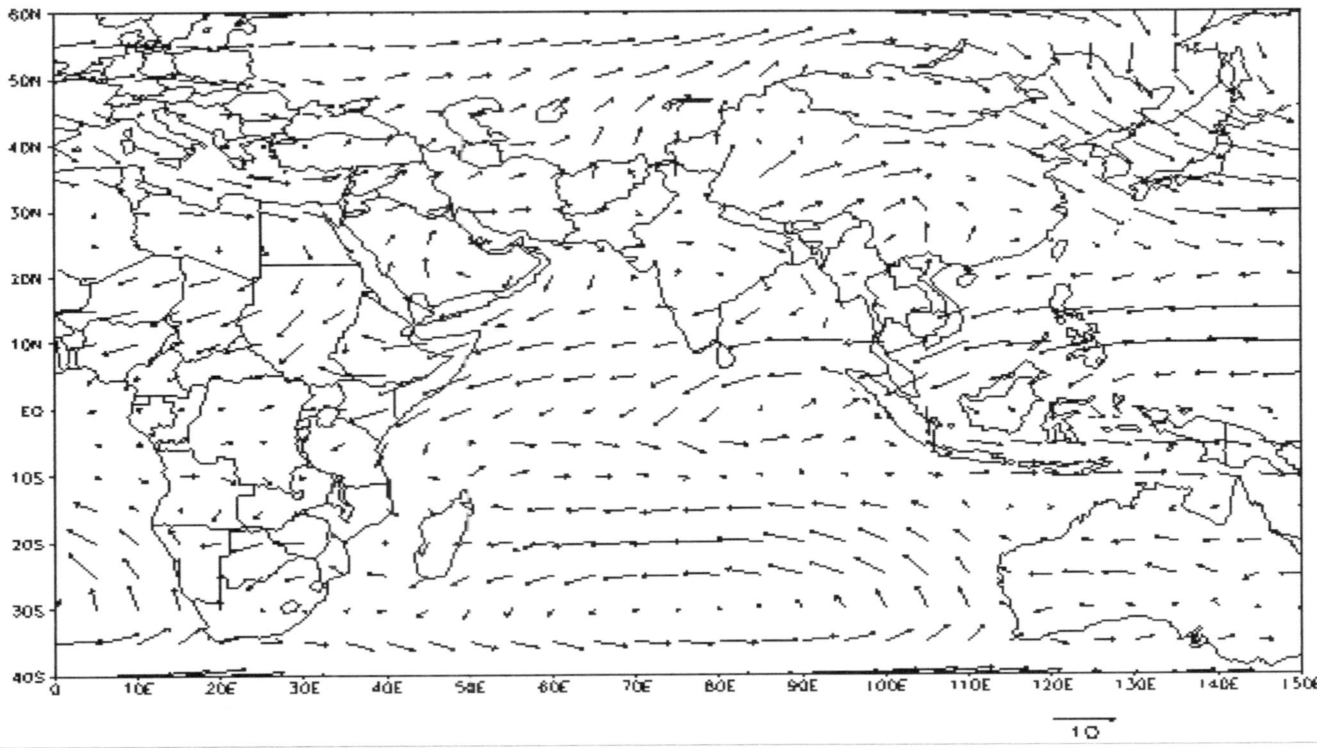

Figure 1.3.5 850 hPa vector wind in January averaged over the period 1979-1995. Wind speed scale in m/sec. (Source: IMD)

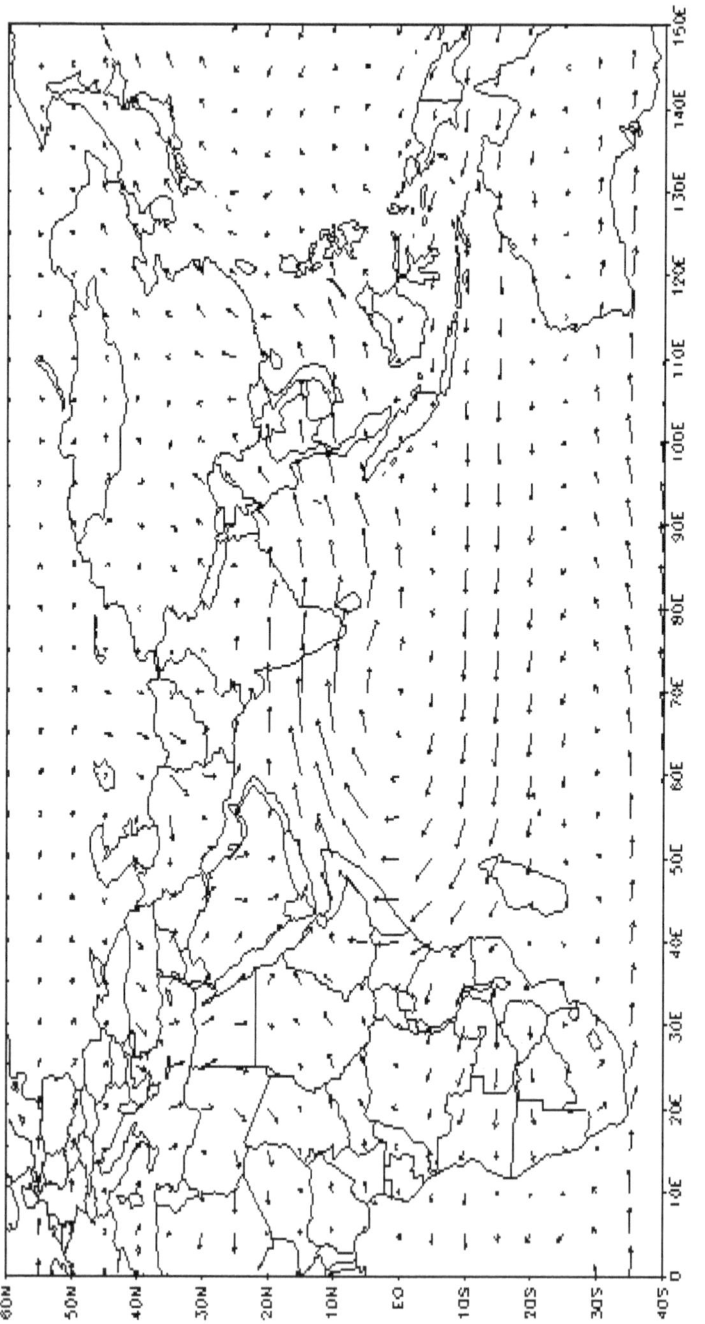

Figure 1.3.6 850 hPa vector wind in July averaged over the period 1979-1995. Wind speed scale in m/sec. (Source: IMD)

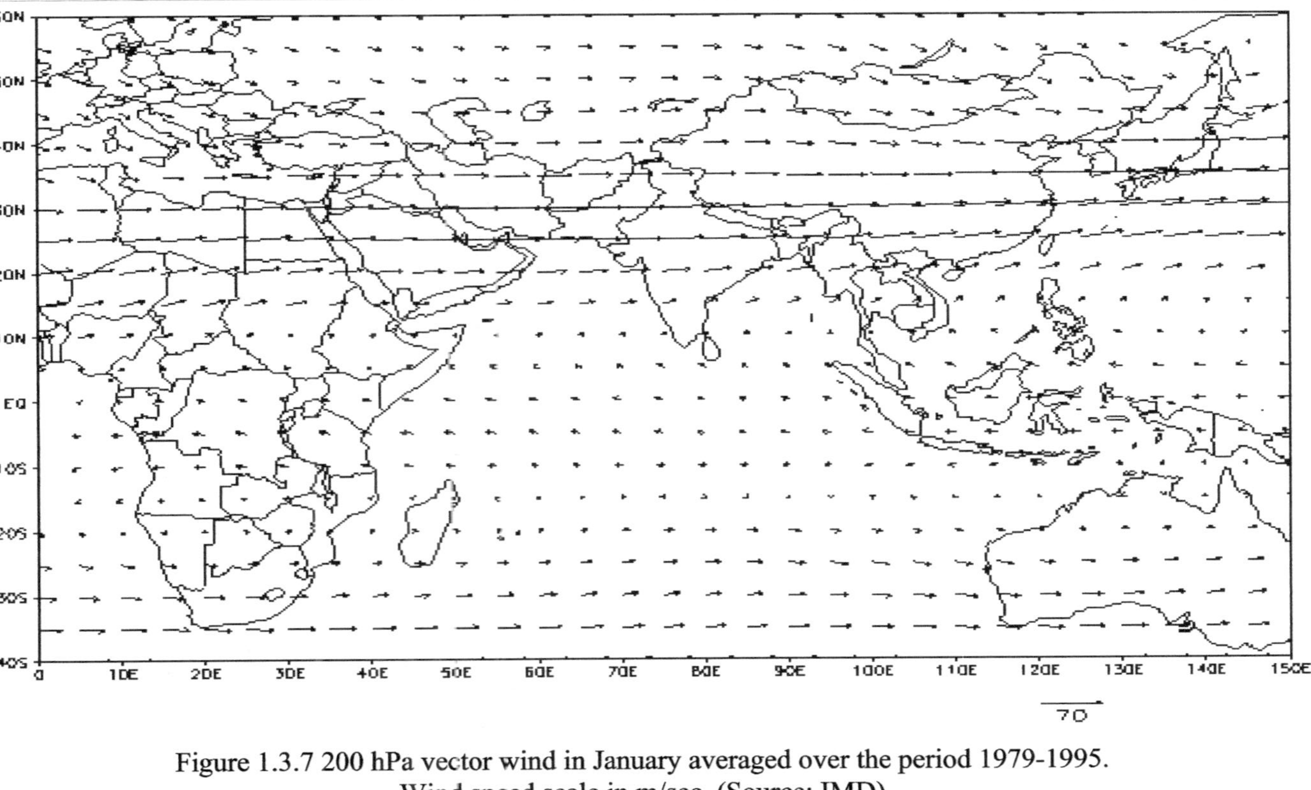

Figure 1.3.7 200 hPa vector wind in January averaged over the period 1979-1995.
Wind speed scale in m/sec. (Source: IMD)

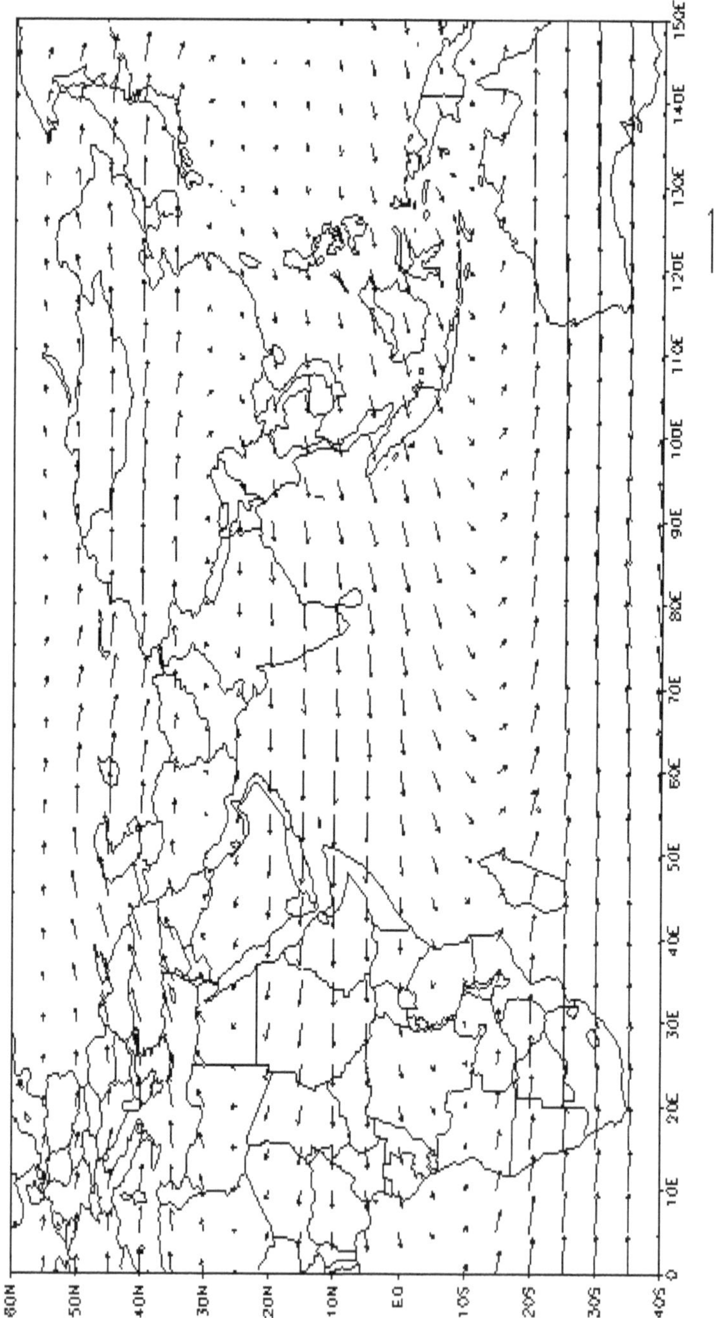

Figure 1.3.8 200 hPa vector wind in July averaged over the period 1979-1995. Wind speed scale in m/sec. (Source: IMD)

## 1.4 Onset and Withdrawal of the Southwest Monsoon

The rainfall over India during the monsoon months of June to September accounts for almost 75-80% of the annual rainfall. Over many parts of the country, the rainfall in the remaining eight months is almost negligible in comparison. Therefore, the entire Indian population eagerly looks forward to the commencement of the monsoon rainfall, especially for a respite from the scorching summer that precedes it. Farmers, in particular, have to carry out their sowings so as to synchronize them with the monsoon rains, else they run the risk of their efforts being rendered futile. The onset of the southwest monsoon over Kerala paves the way for its further advance into the interior parts of the country and it is therefore important not only for Kerala but for entire India. Kerala is India's gateway to the southwest monsoon.

From the meteorological viewpoint, the onset of the monsoon over Kerala is a part of a large scale circulation event. It is associated with changes in many atmospheric and oceanic parameters that might have been building up gradually or which may occur very suddenly. Not only rainfall, but several other factors such as low level and upper level winds, moisture, land and sea surface temperature are involved in the onset process.

The chain of events that culminate in the onset of the monsoon over Kerala is initiated by the low level crossequatorial flow off the coast of Sumatra in early May. This results in an increased moisture flux over the Arabian Sea, deep cumulus convection, latent heat release and a rise in tropospheric temperatures. When these processes persist for some time there is a positive moisture feedback to the atmosphere, a strengthening of the tropospheric westerlies and the development of the Somali low level jet.

In early April, the region of the tropical north Indian Ocean and adjoining west Pacific Ocean begins to warm rapidly, and by the end of May, it becomes the warmest tropical ocean area. The monsoon makes its appearance over south Bay of Bengal, the Andaman Sea and Myanmar coast in May itself, even before the onset over Kerala. By June, it covers the whole Bay of Bengal and Arabian Sea, and extends eastwards into southwest China and the South China Sea (Wang et al 2002, 2004). By the end of June, there is an establishment of a continuous rainbelt that stretches from the Arabian Sea to the South China Sea and beyond.

The mean onset date over Kerala is 1 June with a standard deviation of a week. The monsoon onset over Kerala, however, varies from year to year not only in terms of the date but in its nature. In some onsets, the rainfall may be of the order of 10 cm/day, accompanied by thunder, strong westerly winds, and rough seas. In other years, the onset could be of a very subdued nature.

The earliest ever onset has occurred on 11 May (1918, 1955) and has been as late as on 18 June (1972).

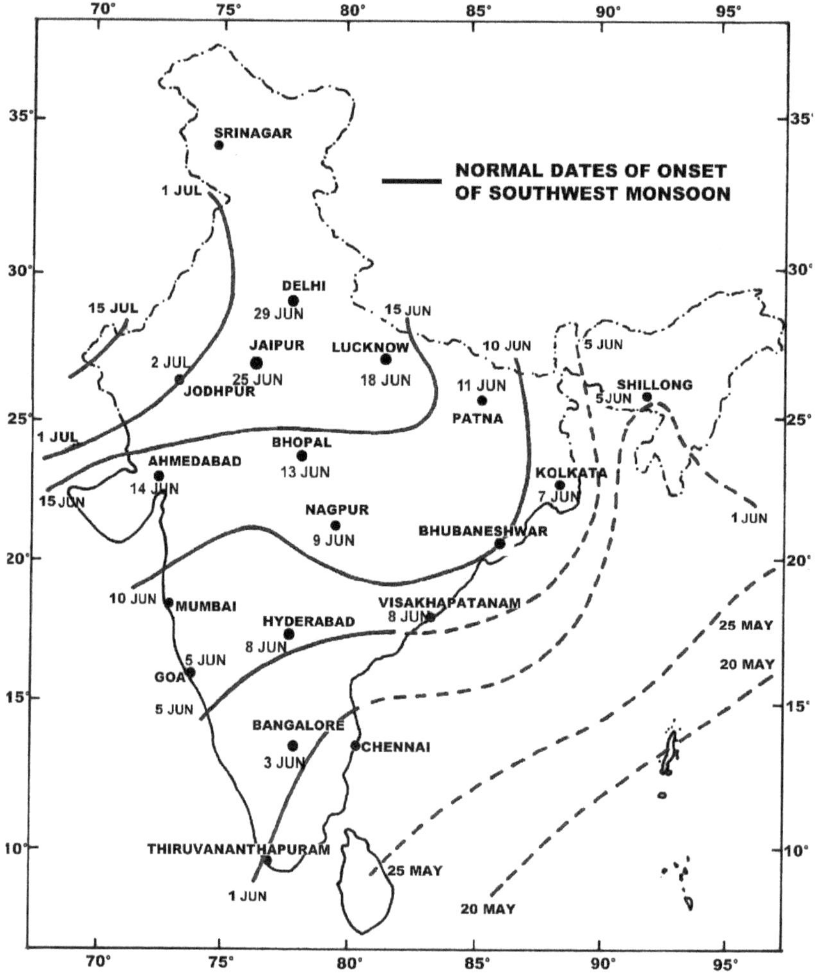


Figure 1.4.1 Normal dates of onset of southwest monsoon (Source: IMD)

The subsequent advance of the monsoon into the interior parts of the country is a slow process and the monsoon covers the entire country by 15 July. Figure 1.4.1 shows the normal dates of arrival of the southwest monsoon over different places. The standard deviation of these onset dates is about 6 to 8 days. In any given year, the actual onset dates are determined by the relative strengths of the Arabian Sea branch and the Bay of Bengal branch of the monsoon. If the Arabian Sea branch is more active, it nudges the monsoon along the west coast into Gujarat and sometimes even into

Rajasthan. If the other branch is stronger, the monsoon gets a push into the eastern parts and then onwards across the northern plains. The northward propagation of the monsoon is governed by different factors and is in the form of a shift of the zone of deep convection, which is seen as the monsoon cloud zone in satellite images (Sikka and Gadgil 1980).

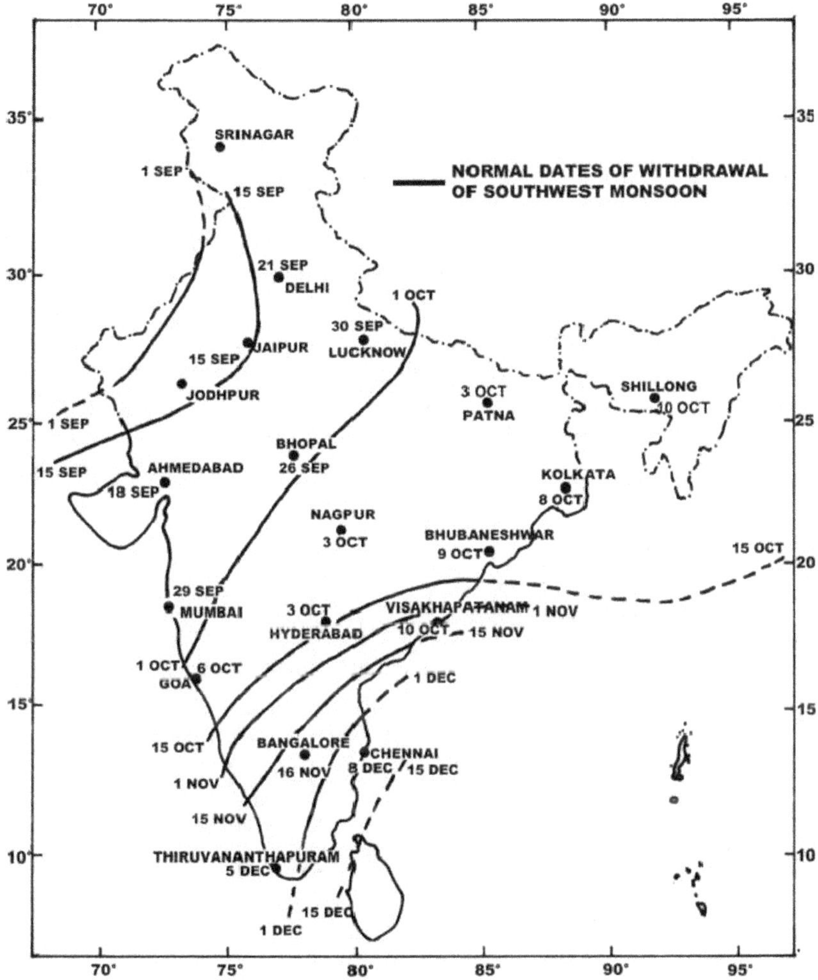


Figure 1.4.2 Normal dates of withdrawal of southwest monsoon
(Source: IMD)

The withdrawal of the southwest monsoon commences from the extreme corners of northwest India by 1 September on an average. The withdrawal process is gradual and completes by 15 October (Figure 1.4.2). However, over the southern peninsula, as the southwest monsoon gives way to the northeast monsoon, rainfall continues until the end of December in association with the northeast monsoon which replaces the southwest monsoon.


## 1.5 Climatology of Monsoon Rainfall

It is most fortunate that India possesses records of rainfall measurements that date back to the 18th century, the British East India Company having established several observatories during its regime. The first rainfall measurement was made at Madras (now Chennai) in September 1793 at the astronomical observatory that had been established there a year before. More observatories were established later, and Simla (now Shimla) and Colaba (in Mumbai) started recording rainfall in 1841. By that time, many provincial governments in India were also making their own measurements of rainfall at a number of stations under the supervision of their revenue officials. However, they used different types of instruments and observational practices and the rainfall data were published in the provincial gazettes. After the establishment of the India Meteorological Department (IMD) in 1875, a great deal of uniformity was brought into the rainfall measurements across the country and by 1889, IMD had started compiling all-India rainfall statistics (IMD 1975).

IMD's role in the collection of rainfall data across the country was accorded a formal status in August 1890 under a resolution of the Government of India, which later came to be known as the 'Resolution on Rainfall Registration in India'. This move of the government made it mandatory for all rainfall measurements in India to be made by means of identical rain gauges of the Symon design, and for the collected data to be sent to IMD regularly. IMD was assigned the responsibility of calibrating all the instruments, and the scrutiny and publication of all rainfall data for India.

During the $20^{th}$ century, regular rainfall measurements were started at thousands of places in India by many different agencies like state revenue departments, railways, agricultural institutions, tea plantations, reservoir management boards, health authorities and others, primarily for their own specific needs and purposes. Even today, the non-IMD rainfall stations far outnumber the IMD observatories. However, as per the requirements of the Resolution on Rainfall Registration, all such rainfall data are received at IMD for scrutiny and compilation of rainfall statistics for India. The records have been meticulously preserved and systematically maintained at IMD's National Data Centre at Pune. This data archive is a gold mine of information on Indian rainfall.

The incorporation of non-IMD rainfall data for the compilation of authentic national rainfall statistics presents many problems that may not be very apparent to the users. The data need to be subjected to very rigid quality checks to eliminate spurious values recorded with poorly maintained instruments or improper observational practices. The raingauges having been set up by agencies primarily for meeting their own requirements, are not distributed uniformly across the country. There are regions where the gauges are clustered together and there are vast areas without a single gauge. Many agencies have not been regular in supplying their data to IMD, producing large gaps in the data series.

For any investigation of the monsoon rainfall over India, a basic requirement is a reliable, quality-controlled, continuous data series. The monthly and annual normals of rainfall and the number of rainy days for all rain-recording stations in India, compiled for the period 1901-1950 by IMD served this purpose for a long time to come. In the 1980's, scientists of the Indian Institute of Tropical Meteorology, Pune brought out a new and updated rainfall data set for the country (Parthasarathy et al 1987) which satisfied the prime requirement of continuity of data. This rainfall data series was derived from the data of 306 raingauge stations that were uniformly distributed across the country, and which had an unbroken record of operation during the period 1871 to 1984. Parthasarathy et al also used this data to compute the area-weighted All-India Summer Monsoon Rainfall (AISMR), but they did not consider the stations in the hilly areas and islands for this analysis. Compared to the massive archives of IMD, the IITM data set was quite limited, but it was user friendly and easy to handle. It has since been used extensively by monsoon researchers both within and outside India in numerous studies of the Indian monsoon rainfall, particularly of its long term variability.

In a very significant effort, IMD has recently compiled a new homogeneous rainfall data set spanning a time period longer than a century, from 1901 to 2003, over an extensive and dense network of 1476 rain gauges across the country (Guhathakurta et al (2007). The analysis includes stations in the hilly regions, and excludes stations which had a break of more than 10% in the data series. The network is so chosen that there are at least two stations in each of the 458 meteorological districts of the country, so that districtwise rainfall statistics can be generated.

In another parallel work, IMD has also produced a high resolution ($1° \times 1°$ lat./long.) daily gridded rainfall data set (Rajeevan et al 2006). This is based upon 1803 stations which had a minimum 90% of data availability during the analysis period of 1951-2003. The new IMD gridded rainfall analysis has a more accurate representation of rainfall over the Indian region, especially along the west coast and northeast India. This data set is expected to be extremely useful for purposes of validation of numerical and climate prediction models.

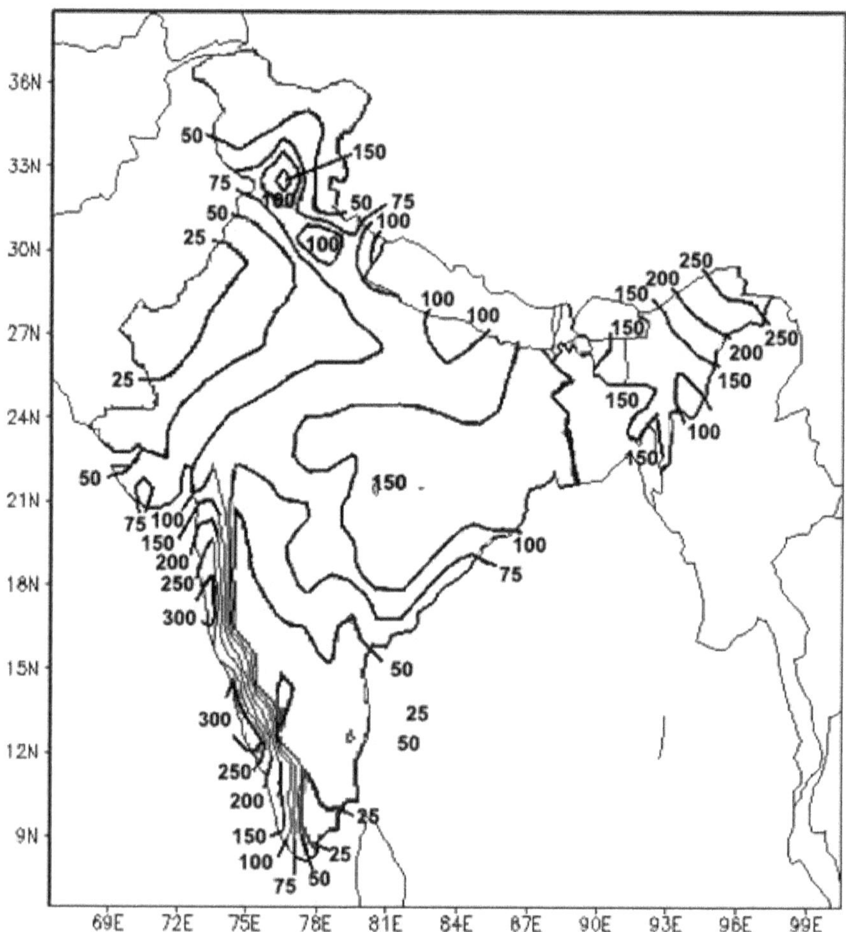

Figure 1.5.1 Normal rainfall (cm) over India during the monsoon season
1 June to 30 September (Source: IMD)

The spatial distribution of the mean seasonal monsoon rainfall (June-September) is depicted in Figure 1.5.1. What is immediately apparent is the wide disparity in the distribution of monsoon rainfall over different parts of the country. There are areas such as Rajasthan in northwest India and adjoining Saurashtra and Kutch, which suffer from perennial dryness, and in some pockets receive barely 10 cm rain in the whole of the monsoon season. The northernmost parts of India get much of their annual rainfall during winter in association with extratropical systems. Similarly over parts of the southern peninsula, the southwest monsoon rainfall is much less than that during the dominant northeast monsoon. The coastal areas windward of the Western Ghats receive the maximum rainfall while in contrast those to the leeward side fall under the rain shadow. The hilly regions of northeast India constitute another maximum rainfall zone. Cherrapunji is well-known as the wettest place on earth, with its normal annual rainfall exceeding 1100 cm.

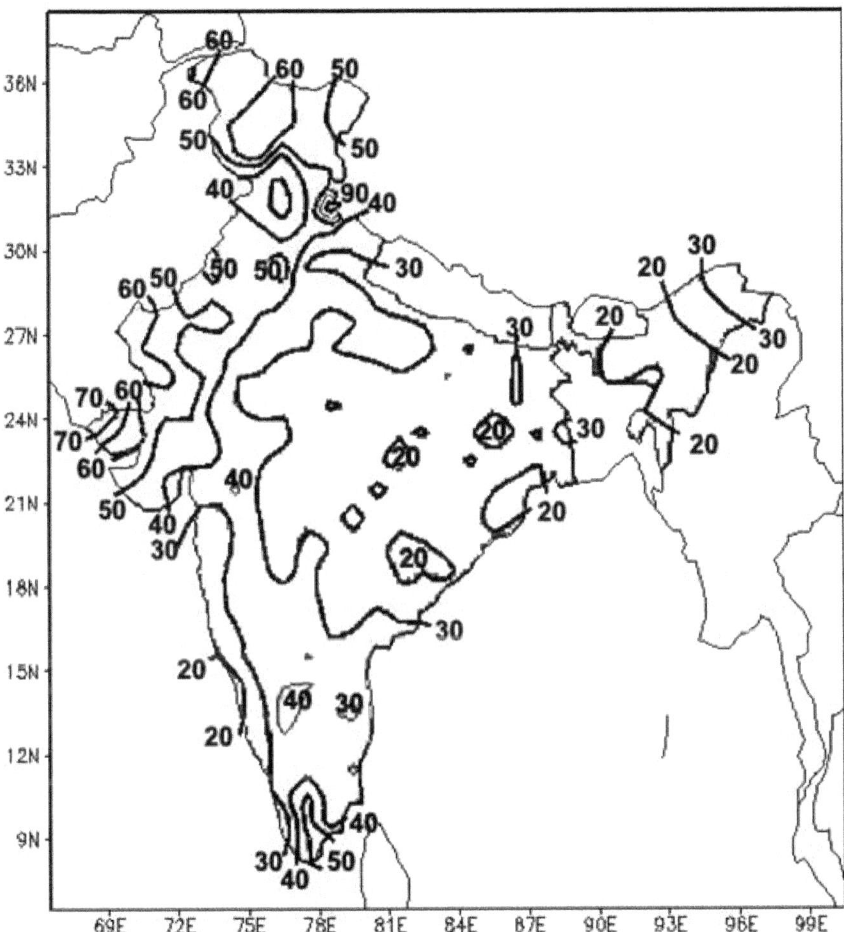


Figure 1.5.2 Coefficient of variation (%) of rainfall over India during the monsoon season 1 June to 30 September (Source: IMD)

The rainfall during the monsoon season is to be viewed together with its coefficient of variation which is defined as the ratio of the standard deviation to the mean expressed as a percentage. The coefficient of variation is of the order of 30% over large areas of the country (Figure 1.5.2). It is 40-50% or higher in regions like Saurashtra, Kutch, Rajasthan and Jammu and Kashmir, which means that their mean rainfall is made up from very heavy rainfall in some years and deficient rainfall in other years. A comparison of Figures 1.5.1 and 1.5.2 reveals that generally speaking, areas having the lowest mean rainfall are also those in which the rainfall varies the most from year to year, while where rainfall is the highest, it is also most regular and assured.

Traditionally, India has been divided into what are known as meteorological subdivisions, which have a common weather and rainfall pattern within themselves, and are administratively convenient as they are drawn along state and internal boundaries. Currently there are 36 meteorological subdivisions in the country (Figure 1.5.3). Some of the states in India are so large that they have been divided into 3 or 4 subdivisions. For example, the state of Andhra Pradesh has three subdivisions: Coastal Andhra Pradesh, Telangana and Rayalaseema, which are climatically different. On the other hand, some of the states are so small that adjacent states which have a similar climate are included in a single subdivision. For example, in the northeast, Nagaland, Mizoram, Manipur and Tripura form one meteorological subdivision.

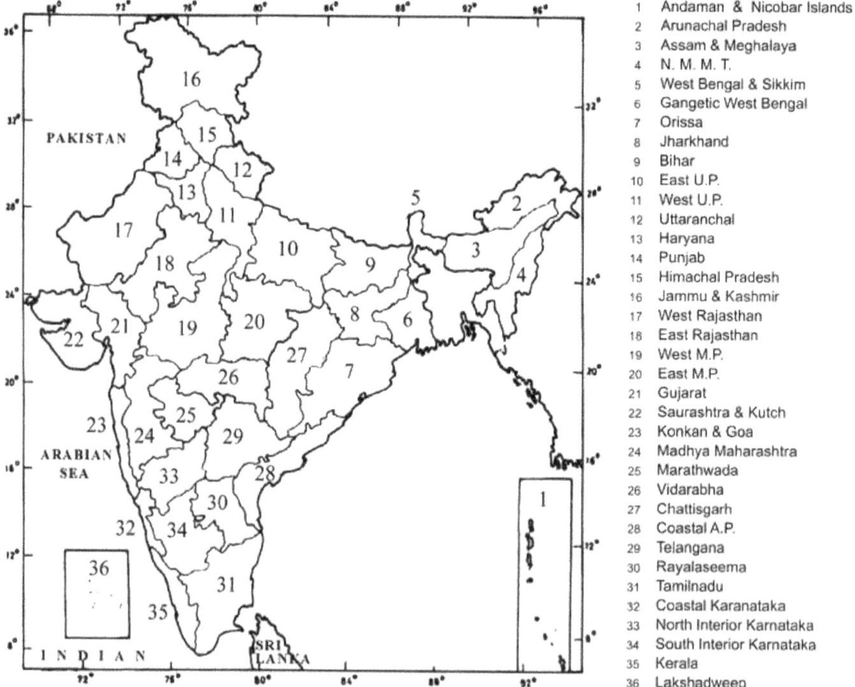


Figure 1.5.3 The 36 meteorological subdivisions of India (Source: IMD)

The rainfall over a subdivision is categorized as excess (>20%), normal (+19 to -19%), deficient (-20 to -59%) or scanty (<-60%) according to the rainfall departure from normal. Rainfall statistics are first generated on the level of the subdivisions and the mean rainfall over the country as a whole is then derived on daily, weekly, monthly, seasonal and annual time scales as the area-weighted rainfall of all the 36 meteorological subdivisions.

Table 1.5.1 and Figure 1.5.4 give the monthwise values of the all-India mean rainfall amounts as well as their standard deviation and coefficient of variation. The months of July and August make the maximum contributions to the all-India annual rainfall of 24.2% (286.7 mm) and 21.6% (255.3 mm) respectively. The other two monsoon months of June and September make a much lower contribution of 13.8% (163.4 mm) and 14.5% (171.8 mm) respectively to the annual rainfall. The all-India mean rainfall for the southwest monsoon season comprising the four months of June to September is 877.2 mm which is as much as 74.2% of the annual rainfall of 1182.8 mm.

**Table 1.5.1 Monthly, seasonal and annual rainfall statistics for India
(Data Source: Guhathakurta et al 2006)**

| Month/Period | Mean Rainfall (mm) | Standard Deviation (mm) | Coefficient of Variation (%) |
|---|---|---|---|
| January | 20.3 | 8.5 | 41.8 |
| February | 24.6 | 10.0 | 40.4 |
| March | 32.0 | 9.2 | 28.8 |
| April | 39.8 | 7.9 | 19.9 |
| May | 61.9 | 12.5 | 20.2 |
| June | 163.4 | 29.5 | 18.1 |
| July | 286.7 | 35.3 | 12.3 |
| August | 255.3 | 30.6 | 12.0 |
| September | 171.8 | 32.8 | 19.1 |
| October | 78.4 | 24.8 | 31.6 |
| November | 30.7 | 15.1 | 49.3 |
| December | 17.9 | 7.9 | 44.0 |
| January-February | 44.9 | 12.3 | 27.3 |
| March-May | 133.7 | 17.7 | 13.2 |
| June-September | 877.2 | 71.0 | 8.1 |
| October-December | 126.9 | 28.8 | 22.7 |
| Annual | 1182.8 | 87.0 | 7.4 |

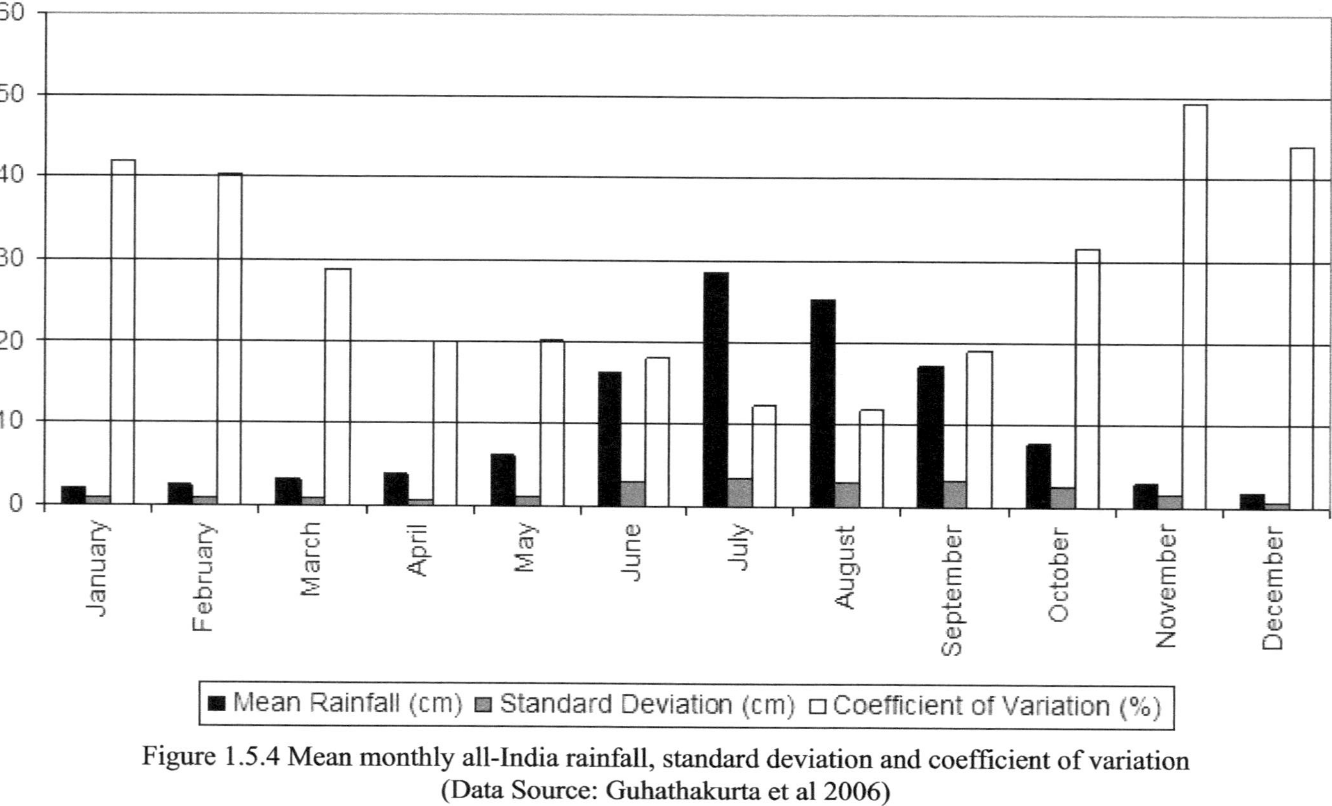


Figure 1.5.4 Mean monthly all-India rainfall, standard deviation and coefficient of variation (Data Source: Guhathakurta et al 2006)

While India receives abundant rainfall compared to that over many other regions of the world including other monsoonal regimes, three-fourths of it falls during the four months of the southwest monsoon season. This makes it necessary to maximize the use of the available water during the monsoon, and to make adequate arrangements for its storage and use during the other eight months of the year.

Another important feature of the monsoon is that the intraseasonal variability of rainfall distribution is much larger than interannual variability. Even in the worst famine that India experienced in the year 1899, the total annual rainfall was 26 percent below normal. In the year 1917, which was an opposite extreme, the rainfall was 29 percent above normal.

On the other hand, rainfall variations of a much higher order occur during the monsoon season. Monsoon rainfall is not continuous but occurs in spells lasting for about 5-7 days which are interspersed with dry spells. During the most active monsoon months of July and August, the monsoon trough runs across the Indo-Gangetic Plains producing good rainfall in the plains of India but when it shifts to the Himalayan foothills, there is subdued rainfall over most parts of the country and if this situation is prolonged it results in what is called a break In the monsoon. Subseasonal variations are also produced by the timing of the onset of the monsoon and its advance into the interior, as well as the timing of its withdrawal.

## 1.6 The Monsoon and Indian Agriculture

There is evidence to show that India had to suffer from recurrent famines throughout its history, and all the more so during the period of British rule. Famines frequently affected many large regions and in some years even the entire country, bringing misery and death to millions of people. Although famines were primarily caused by a failure of the crops due to extremely poor monsoon rains, the difficulties of the population got compounded by the general apathy of the rulers, lack of relief provisions, or inconsiderate measures like increased taxation being introduced at the same time.

The infamous Bengal famine of 1770 was the first to have occurred under the regime of the British East India Company, and it is said to have resulted in the starvation death of a population of ten million in that province. During the nineteenth century, famines kept on affecting different parts of the country, and millions of people died of hunger, since they had no alternative sources of livelihood. Bengal was struck by famine again in 1866 and at the same time Orissa was also badly affected. However, this calamity also had a positive outcome. The commission of enquiry appointed by the British government to examine the circumstances of the famine recommended the setting up of an all-India meteorological organization. This far-reaching recommendation, which was supported by other bodies like the Asiatic Society of Bengal, eventually led to the establishment of the India Meteorological Department in 1875 (IMD 1975).

In the following year, there was a famine in Madras, and there was a countrywide famine in 1899. The year 1918 is remembered as the year of the great Indian famine. In 1943, there was yet another major famine in Bengal leading to the death of at least three million people. India's British rulers were at that time preoccupied with the Second World War, and left the Indian farmers to fend for themselves. Whatever they had harvested was also acquired by the government in the name of the war effort, and grain trading was banned.

Even after India won its struggle for independence in 1947, it had to carry on for another twenty years with its battle on the food front. With the partition of the country, it had lost some of its most fertile land and the population had been increasing continuously. In the decade of the 1960s, India's imports of food grain had shot up to over 10 million tonnes per year. But then came the Green Revolution and it transformed the country into a self-sufficient nation. The building of huge buffer stocks, a good public distribution system, and an efficient relief and disaster management organisation, have all freed the Indian population from the miseries of famine (GNI 2008). The statistics in Table 1.6.1 speak for themselves.

**Table 1.6.1 India's food grain production, imports, buffer stock and population 1950-2000 (Data Source: GNI 2008)**

|  | 1950 | 1960 | 1970 | 1980 | 1990 | 2000 |
|---|---|---|---|---|---|---|
| **Food grain production (million tonnes)** | 50.8 | 82.0 | 108.4 | 129.6 | 176.4 | 201.8 |
| **Food grain import (million tonnes)** | 4.8 | 10.4 | 7.5 | 0.8 | 0.3 | - |
| **Buffer stock (million tonnes)** | - | 2.0 | - | 15.5 | 20.8 | 40.0 |
| **Population (million)** | 361 | 439 | 548 | 683 | 846 | 1000 |

Today, India has a total arable land of 162 million hectares. Out of this, the irrigated area is 52 million hectares, which is the largest amongst all the countries of the world. India is currently the world's largest producer of tea, milk, pulses and jute, the world's second largest producer of wheat, rice, groundnut, vegetables, fruits and sugarcane, and the third largest producer of potatoes and cotton (DAC 2008). Since independence, the Indian agricultural sector has made significant strides in all directions, and the per capita availability of food grains in India has increased considerably in spite of the ever-growing needs of its massive population. However, the monsoons still continue to play a dominating role in Indian agriculture. The meagre rainfall over many regions and topographical features set a limit to the percentage of land that can be irrigated. Furthermore, irrigation water does not have any independent source, it is again the monsoon. Excepting the Himalayan rivers which are fed by snowmelt, all other Indian rivers originate out of the monsoon rain that falls into their catchment areas. The monsoon can thus be said to be India's source of water for agricultural and all other purposes.

The peculiarity of India's annual rainfall is that about 75-80% of it occurs in the monsoon months of June to September. Maximum use of the rain water has to be made in these four months and water stored for use in the remaining eight months of the year. Indian agriculture has traditionally got adjusted with this rainfall pattern. The kharif crop is the rainfed crop that is raised directly on the monsoon rains or with some supplemental irrigation if available, and the rabi crop is raised on the residual soil moisture that the monsoon has left behind after its withdrawal in October. In north India, the occasional winter rains help to boost the crop productivity, particularly in the case of wheat, but elsewhere only those crops that can do without water can be grown in the rabi season. For southern India, especially Tamil Nadu, however, the main agricultural season is associated with the northeast monsoon months of October to December.

The percentage of the irrigated area to the total agricultural land area in India varies widely from state to state (Figure 1.6.1). Only Punjab and Haryana are in a fortunate position with more than 80-90% land under irrigation while Uttar Pradesh, Bihar and West Bengal have a half or more of irrigated land. For most other states, the figures are around 10-30%. As a result, for the country as a whole, only 40% area has the availability of irrigation water and the remaining 60% is completely dependent upon rainfall and it contributes to about a half of the total food grain production. The irrigation potential has largely been achieved over many parts of the country and there is not much scope to bring more land under irrigation in the future.

Although the broad pattern of Indian agriculture is tuned to the annual distribution of rainfall, it is the vagaries of the monsoon and the distribution of monsoon rainfall across the country and within the season, that heavily impact agricultural production. The timeliness of monsoon onset, the timing of active and break phases, the duration of dry and wet spells, the effects of weather on the incidence of pests and diseases, all influence the food grain production that is achieved at the end of the kharif season. The cropping pattern over the country has evolved in consonance with the long term climate prevailing in various parts of the country but the crop acreage and production in any given year depends upon the amount and distribution of rainfall that was actually available on the smaller spatial scales vis-à-vis the water requirements of specific crops at their critical growth stages. Thus drought can occur at the district or subdivision level even when the country as a whole has received a statistically normal rainfall.

The influence of the monsoon rains on the country's agriculture is clearly evident in the graph of the country's food grain production over the 40-year period from 1966-67 to 2006-07 (Figure 1.6.2). It shows a steady rising trend that is attributable to the continuing improvements in technology, farming practices, increase in area sown and availability of irrigation. In fact the food grain production has more than doubled over the last 40 years. However, the graph has peaks that correspond to the exceptionally good monsoons of 1970, 1975 and 1983, and dips that correspond to the all-India droughts of 1966, 1972, 1979, 1982, 1987 and 2002. One of the worst and most recent examples of monsoon failure is the agricultural year 2002-03, in which food grain production fell to 175 million tonnes from the previous year's record production of 213 million tonnes. This record of 2001-02 remained unsurpassed even though the production recovered again in 2003-04 after a good monsoon. Another noticeable feature of this graph is the rise in rabi production to the extent that it is now almost catching up with the kharif season crops. In 2002-03, for the first time, the rabi food grain production of 87.5 million tonnes had even slightly exceeded the kharif production figure (DAC 2008).

Besides directly impacting the agricultural production of the country, large scale deficiencies of monsoon rainfall cause many other problems such as shortage of fodder for animals and scarcity of drinking water, while excess rains lead to flooding, disruption of normal life and loss of standing crops. During the monsoon season, crops are also lost to weather-related pests and diseases.

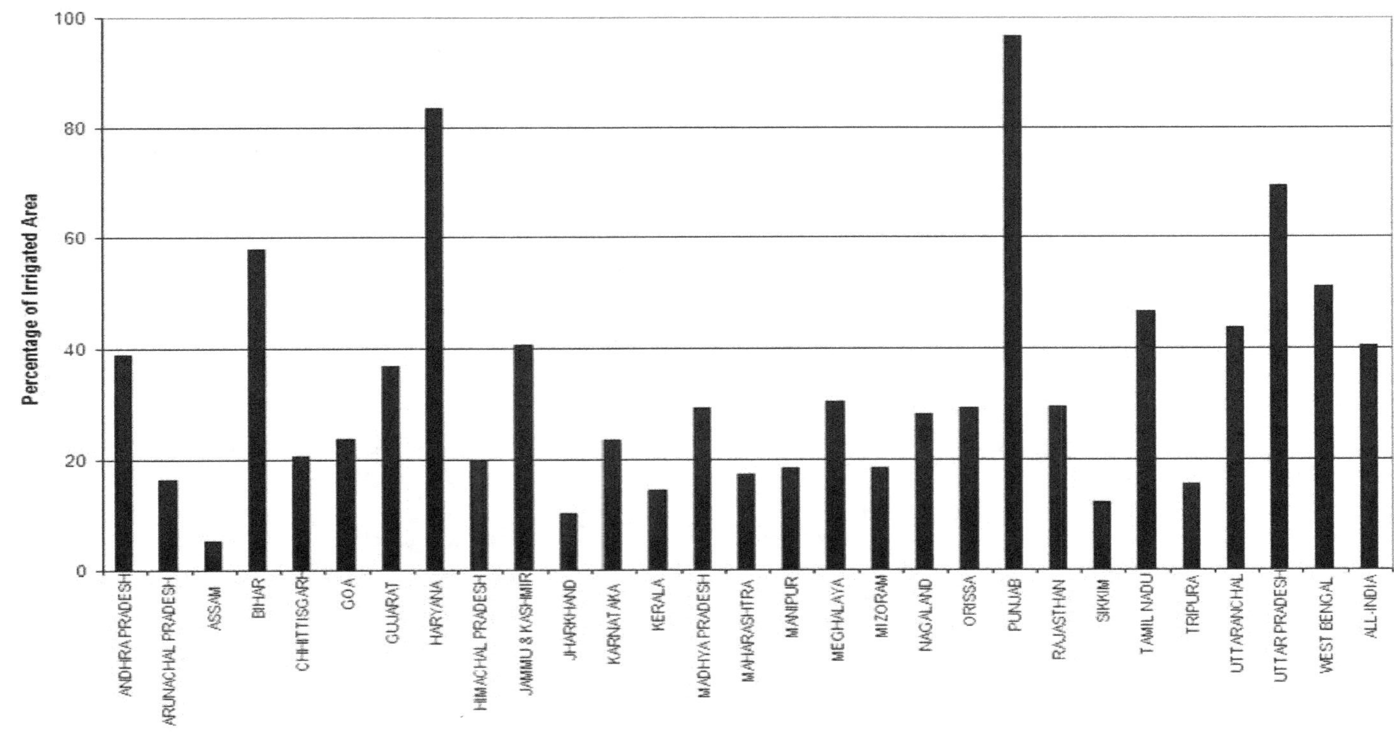


Figure 1.6.1 Percentage of irrigated area of principal crops in various states of India in 2003-04 (Source: DAC 2008)

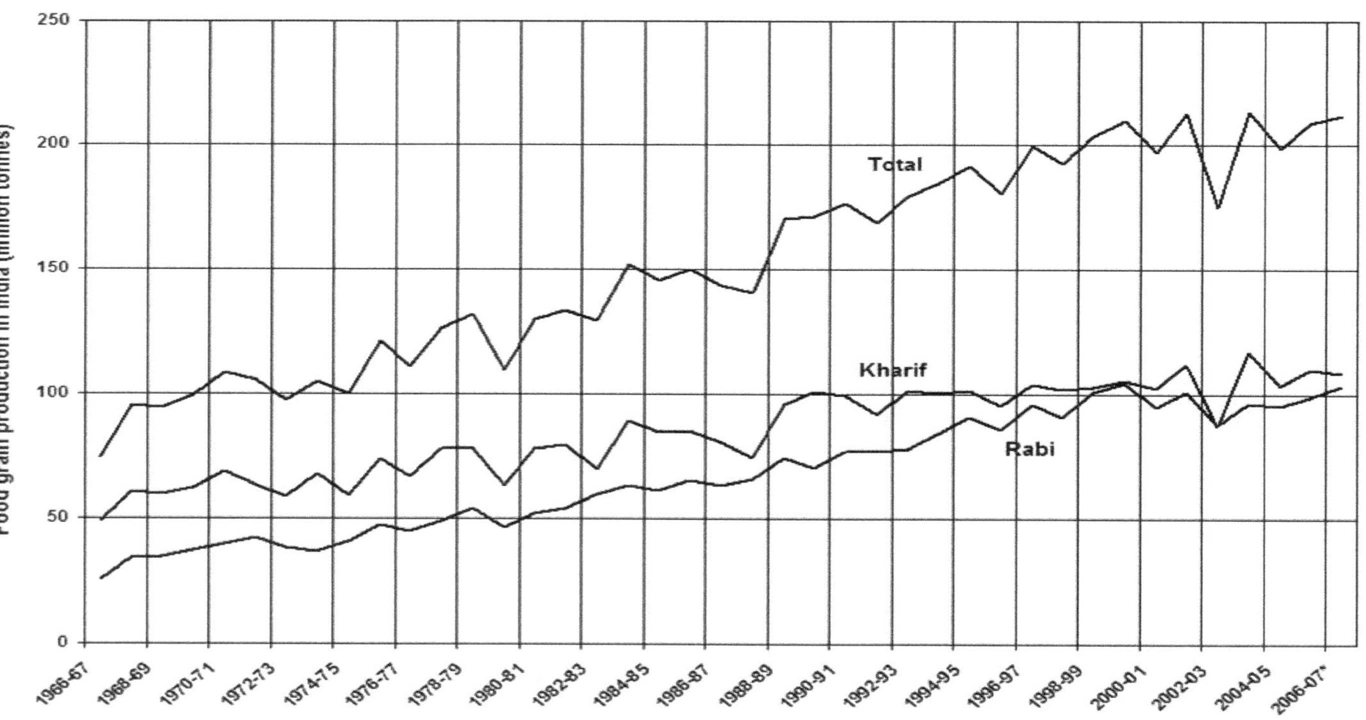


Figure 1.6.2 Food grain production in India during kharif and rabi seasons and annual total, from 1966-67 to 2006-07. *Estimate for 2006-07 is tentative. (Source: DAC 2008)

## 1.7 Impact of the Monsoon on the Indian Economy

The vulnerability of the Indian economy to the vagaries of the monsoon has reduced over recent years because of the emergence of new growth drivers such as software, telecom, pharmaceuticals and other sectors. Even then, it cannot be emphatically said that the monsoon is no longer of any consequence. More than half of India's population still derives its income from agriculture, and the larger market for consumer goods is in the rural areas of the country. So a failure of agriculture on account of the monsoon does reduce the purchasing power of the farming community, while a good monsoon does generate a greater demand for consumer products.

After the drought of 2002, there were fears of another consecutive poor monsoon year. When the monsoon forecasts dispelled such apprehensions, the Indian economy began to bounce again in 2003 under what came to be known as a 'feel good factor'. Figure 1.7.1 shows the variation in India's GDP growth rate over the years 1997 to 2007 (RBI 2008).

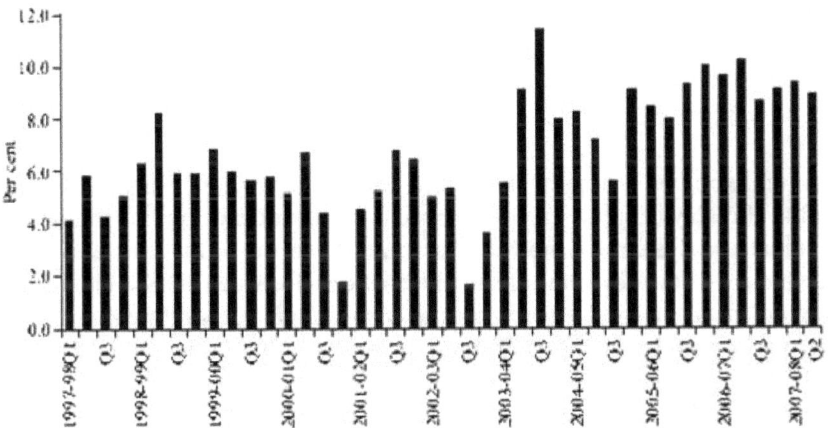

Figure 1.7.1 India's GDP growth rate from 1997 to 2007 (Source: RBI 2008)

The adverse impacts of the drought of 2002 and the deficient monsoon of 2004 on the GDP growth rate are clearly evident in the graph. In 2007, rainfall during the southwest monsoon season was 5 per cent above normal and it was also well-distributed over time. This helped in improving the 2007 kharif sowings and the sown area was 2.7% higher than the previous year. The first advance estimates of the total kharif food grain production are placed at 112.2 million tonnes, about 1.6 per cent higher than that of the previous year (RBI 2008). This improved performance of the agricultural sector is also reflected positively in the GDP growth rate graph. The Indian economy is continuing to exhibit a robust growth and a GDP growth rate of over 9% is being talked about.

## 1.8 Pioneering Work in Monsoon Prediction

No discussion about the prediction of the Indian southwest monsoon can be regarded as complete without a reference to the pioneering efforts of Henry Blanford, John Eliot and Gilbert Walker.

### 1.8.1 Henry Blanford and John Eliot

Since 1867, Henry F. Blanford had been serving as the Meteorological Reporter to the Provincial Government of Bengal. When IMD was established in 1875, he was asked to take charge as Imperial Meteorological Reporter to the Government of India. In 1877 the monsoon rains had failed and the Government was anxious to know the prospects of the monsoon in 1878. John Eliot, who was officiating in place of Blanford at that time, examined the previous meteorological statistics and submitted an opinion that the 1878 rainfall would be more equitably distributed than in 1877, a forecast which came quite correct (IMD 1975).

Blanford, who had a long acquaintance with Indian weather, had postulated that the Himalayan snow cover exercised a great influence on Indian rainfall. He initiated a system of snowfall observations in the Himalayas and made arrangements for obtaining this information. Between 1881 and 1885, Blanford made his first experiments at making long range forecasts of the monsoon rainfall on the basis of the snow reports. On 4 June 1886, he issued the first long range monsoon forecast of IMD, which was published in the Gazette of India. India thus became the first country in the world to issue seasonal forecasts on an operational basis.

Blanford retired in 1889 and was succeeded by John Eliot who was given a new designation of Director General of Observatories. Eliot continued the practice of issuing long range forecasts, adding a memorandum that described the general meteorological conditions and abnormal features preceding the monsoon. He began to use several predictors in addition to snowfall and he introduced the method of analogues and curve parallels. However, Eliot's forecasts for the years of 1899 and 1901 failed to come true.

## 1.8.2 Gilbert Walker

Amongst all British meteorologists who worked in India, the name of Sir Gilbert Thomas Walker ranks high by any standard. Born on 14 June 1868, Gilbert Walker grew to be a brilliant mathematics scholar at Cambridge, a Senior Wrangler in 1889, and a Fellow of Trinity College in 1891. He worked in many fields, including electromagnetism and the dynamics of projectiles. Walker, however, got chosen by the Government of India to succeed John Eliot as the chief of the India Meteorological Department and he assumed charge as the Director General of Oservatories on 1 January 1904. Soon thereafter he was elected a Fellow of the Royal Society and received the Sc. D. degree from Cambridge University. Walker had the longest tenure of 20 years as the Director General of IMD. In 1918, Walker was elected as the President of the Asiatic Society of Bengal and also of the Indian Science Congress. He was a member of the Board of Governors of the Indian Institute of Science, Bangalore. He was knighted in 1924, the year of his retirement. Sir Gilbert, on return to England, worked as Professor of Meteorology at the Imperial College of Science and Technology, London. Here, he continued his research on varied subjects like cloud formation, convection in unstable fluids and the flight of birds. He was the President of the Royal Meteorological Society in 1926-27. He died at the age of 90, on 4 November 1958.

Sir Gilbert Walker, the mathematician-turned-meteorologist, lived decades ahead of his times. The phenomenon of the southern oscillation which together with the El Nino is now popularly known as ENSO, and which is monitored with the help of satellites, was in fact discovered by Walker at a time when meteorological and oceangraphic observations were sparse and made only at the surface. In an era in which the means of acquiring and processing such data were the most primitive by today's standards, Walker could establish the fact that the Indian monsoon was not an isolated system but had strong teleconnections with the global climate. He was the first to apply statistical techniques to the foreshadowing of Indian monsoon rainfall using antecedent parameters measured in different parts of the world. Walker's work was so robust that eighty years after he left India, his statistical approach continues to be used for long range forecasting of the Indian southwest monsoon rainfall.

Walker's memory has been perpetuated by the global meteorological community by naming the east-west circulation over the eastern Pacific Ocean as the Walker Circulation. In 2006, the University of Reading established a research institute dedicated to the memory of Sir Gilbert Walker, and named it as the Walker Institute for Climate System Research.

More information about the Walker Institute and about the life of Sir Gilbert Walker is available on the internet (Walker Institute 2008). A collection of Walker's research papers on the Indian monsoon rainfall and its global linkages has been published by the Indian Meteorological Society, New Delhi (IMS 1986) and a small selection is available online (RmetS 2008).

## 1.9 The Need for Monsoon Prediction

In the 1880s when India was being ruled by the British, the government had desired to know in advance about the likely quantum of monsoon rainfall. British meteorologists working in India had responded to that call. Even in today's independent India with its thriving economy, a reliable monsoon prediction is in no way of a lesser importance.

From the facts and figures quoted in the preceding sections, it can be concluded without any doubt that the monsoon continues to have a hold on Indian agriculture and through it, on the Indian economy as a whole. The crop productivity of some of the rainfed zones is as low as 1 tonne/hectare, which is a sad reflection of the agricultural risks that go with the vagaries of the monsoon. Monsoon prediction is therefore of great value in any efforts aimed at minimizing the agricultural risks and maximising the crop yields. However, a single common prediction cannot serve all purposes. Different types of predictions of the behaviour of the monsoon are required to be generated on appropriate time and space scales, to suit applications ranging from national and state level policy making, to managing crops in farmers' fields. Long term food security concerns have also to be viewed against the impending threat of global warming and it is necessary to assess its likely impact on the amount and distribution of the monsoon rainfall.

## 1.10 References

Adams D. K. and Comrie A. C., 1997, "The North American Monsoon", *Bulletin American Meteorological Society*, 78, 2197-2213.

AMMA, 2008, African Monsoon Multidisciplinary Analysis web site http://amma-international.org.

Arkin P. A., Rao A. V. R. K. and Kelkar R. R., 1989, "Large-scale precipitation and outgoing longwave radiation from INSAT-1B during the 1986 southwest monsoon season", *J. Climate*, 2, 619-628.

Asnani G. C., 1993, *Tropical Meteorology*, Vol. 1-2, 1202 pp.

Asnani G. C., 2005a, "ITCZ (Inter-Tropical Convergence Zone)", *Tropical Meteorology*, Vol. 1, Chapter 2, 126-144.

Asnani G. C., 2005b, "Special features of tropical meteorology", *Tropical Meteorology*, Vol. 1, Chapter 1, 1-121.

Asnani G. C., 2005c, "Physics and dynamics of monsoon", *Tropical Meteorology*, Vol. 1, Chapter 4, 3-11.

Asnani G. C., 2005d, "Monsoon of South Asia, Part II: Tibetan Plateau, Lowlands of China, Indochina peninsula, Philippines and Japan", *Tropical Meteorology*, Vol. 3, Chapter 12, 1-130.

Asnani G. C., 2005e, "Monsoon of Australia, Indonesia, Fiji Islands and New Zealand", *Tropical Meteorology*, Vol. 3, Chapter 13, 1-67.

Asnani G. C., 2005f, "Monsoon of north, central and south America", *Tropical Meteorology*, Vol. 3, Chapter 15, 1-52.

Asnani G. C., 2005g, "Monsoon of Africa", *Tropical Meteorology*, Vol. 3, Chapter 14, 1-106.

Asnani G. C., 2005h, "Monsoon of South Asia, Part I: Southwest Asia, India and neighbourhood", *Tropical Meteorology*, Vol. 3, Chapter 11, 1-248.

DAC, 2008, Department of Agriculture and Cooperation, Ministry of Agriculture, Government of India, New Delhi, web sites http://agricoop.nic.in/ Agristatistics.htm and http://dacnet.nic.in/eands/agStat06-07.htm.

Douglas M. W. and coauthors, 1993, "The Mexican monsoon", *J. Climate*, 6, 1665-1677.

Gadgil S., 2007, "The Indian monsoon - 3. Physics of the monsoon", *Resonance*, 12, 5, 4-20.

GNI, 2008, "The green revolution in India", Good News India web site http://goodnewsindia.com.

Guhathakurta P. and Rajeevan M., 2006, "Trends in the rainfall pattern over India", *Research Report No. 2/2006*, National Climate Centre, India Meteorological Department, Pune, 25 pp.

Guhathakurta P. and Rajeevan M., 2007, "Trends in the rainfall pattern over India", *Int. J. Climatology*, DOI: 10.1002/joc.1640.

Halley E., 1686, "An historical account of the trade winds and monsoons, observable in the seas between and near the tropics with an attempt to assign the physical cause of the said winds", *Philosophical Transactions, Royal Society*, London, 16, 153-168.

Hendon H. H. and Liebmann B., 1990, "A composite study of the onset of the Australian summer monsoon", *J. Atmospheric Science*, 47, 2227-2240.

Higgins R. W. and co-authors, 2003, "Progress in Pan American CLIVAR research: The North American monsoon system, *Atmosfera*, 16, 29-65.

Holland G. J., 1986, "Interannual variability of the Australian summer monsoon at Darwin: 1952-82", *Monthly Weather Review*, 114, 594-604.

Holland G. J. and coauthors, 1986, "The BMRC Australian Monsoon Experiment: AMEX", *Bulletin American Meteorological Society*, 67, 1466-1472.

IMD, 1975, *Hundred Years of Weather Service (1875-1975)*, India Meteorological Department, Pune, 207 pp.

IMS, 1986, *Sir Gilbert Walker Selected Papers, Long Range Forecasting of Monsoon Rainfall*, Indian Meteorological Society, New Delhi, 254 pp.

Kripalani R. H., Oh J. H. and Chaudhari H. S., 2007, "Response of the East Asian summer monsoon to doubled atmospheric $CO_2$: Coupled climate model simulations and projections under IPCC AR4", *Theoretical Applied Climatology*, 87, 1-28.

Nogues-Paegle J. and coauthors, 2002, "Progress in Pan American CLIVAR research: Understanding the South American monsoon", *Meteorologica*, 27, 3-30.

Parthasarathy B., Sontakke N. A., Munot A. A. and Kothawale D. R., 1987, "Droughts/floods in the summer monsoon rainfall season over different meteorological subdivisions of India for the period 1871-1984", *J. Climatology*, 7, 57-70.

Rajeevan M., 2006, Bhate J., Kale J. D. and Lal B., 2006, "High resolution daily gridded rainfall data for the Indian region: Analysis of break and active monsoon spells", *Current Science*, 91, 296-306.

Ramage C. S., 1971, *Monsoon Meteorology*, Academic Press, New York, 296 pp.

RBI, 2008, Reserve Bank of India, web site http://rbi.org.in.

RMetS, 2008, Royal Meteorological Society web sites
http://www.rmets.org/publication/classics/cp2.php and
http://www.rmets.org/publication/classics/walkerbliss.php.

Sanjeeva Rao P. S. and Sikka D. R., 2007, ""Interactive aspects of the Indian and the African summer monsoon systems", *Pure Applied Geophysics*, 164, 1699-1716.

Sikka D. R. and Gadgil S., 1980, "On the maximum cloud zone and the ITCZ over Indian longitudes during the southwest monsoon", *Monthly Weather Review*, 108, 1840-1853.

Srinivasan J. and Joshi P. C., 2007, "What have we learned about the Indian monsoon from satellite data?", *Current Science*, 93, 165-172.

Trenberth, K. E., Stepaniak D. P. and Caron J. M., 2000, "The global monsoon as seen through the divergent atmospheric circulation", *J. Climate*, 13, 3969-3993.

VAMOS, 2008, Variability of the American Monsoon System web site http://www.clivar.ucar.edu/science/vamos.htm.

Vasisht M. C., 2003, "Influence of monsoon winds", *Kerala Calling*, August 2003, 24-26, http://www.kerala.gov.in/keralacallingaug/p24-26.pdf.

Vera C. and coauthors, 2006, "Towards a unified view of the American monsoon systems", *J. Climate*, 19, 4977-5000.

Walker Institute, 2008, Walker Institute web site http://www.walker-institute.ac.uk.

Wang B. and Lin H., 2002, "Rainy season of the Asian-Pacific summer monsoon", *J. Climate,* 15, 386-398.

Wang B., Lin H., Zhang Y. and Lu M. M., 2004, "Definition of South China Sea monsoon onset and commencement of the east Asia summer monsoon", *J. Climate,* 17, 699-710.

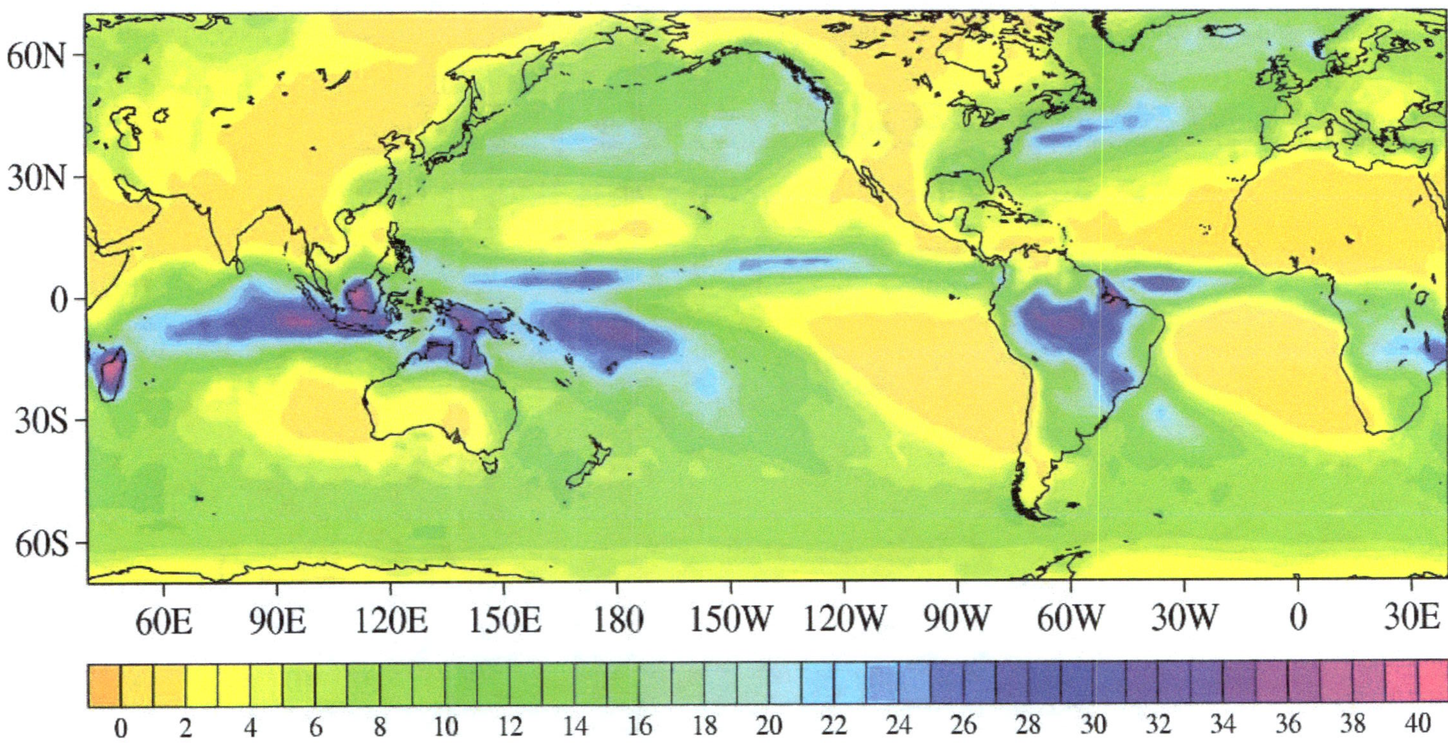



Figure 1.3.1 Global Precipitation Climatology Project (GPCP) precipitation (cm) in January averaged over the period 1979-2007 (Source: http://jisao.washington.edu/data/gpcp)

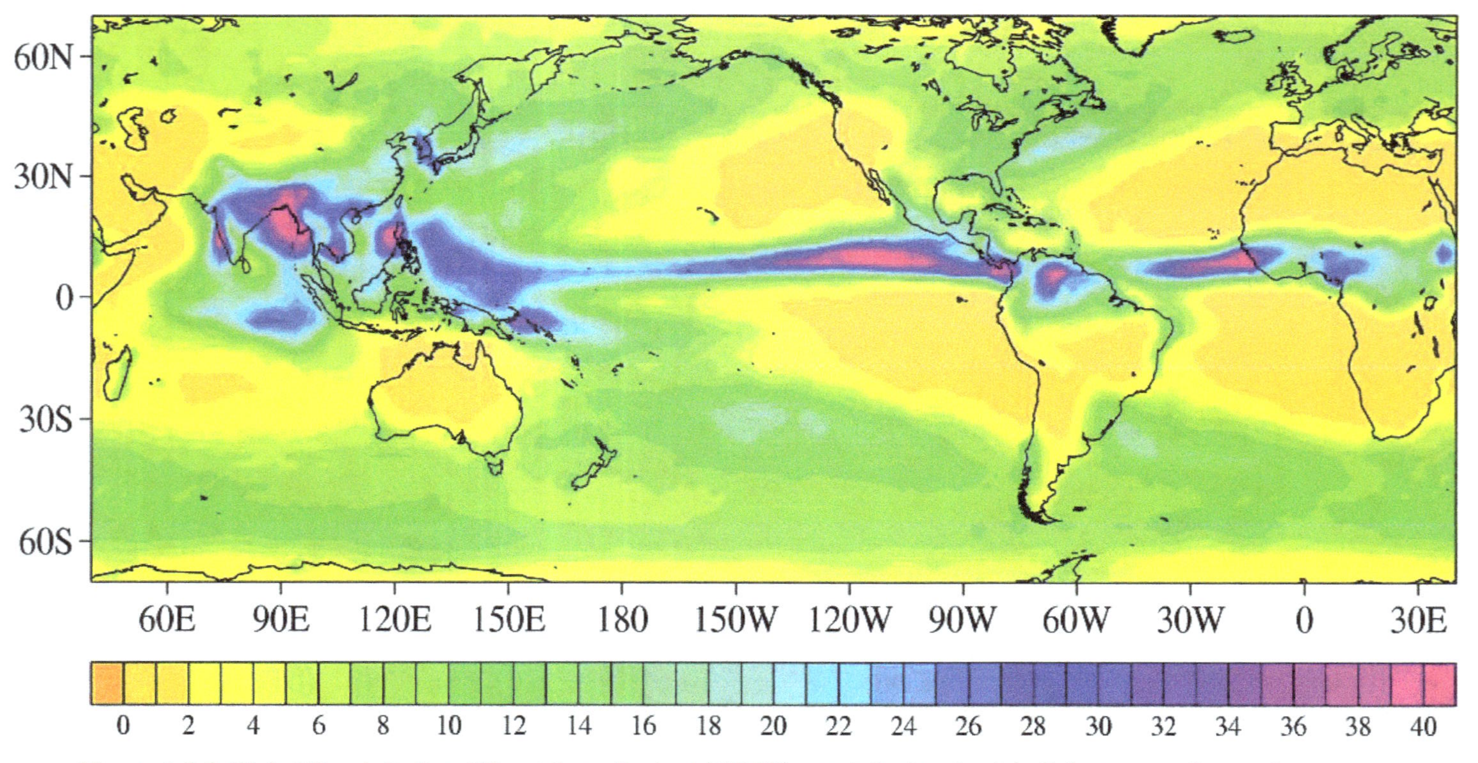


Figure 1.3.2 Global Precipitation Climatology Project (GPCP) precipitation (cm) in July averaged over the period 1979-2007 (Source: http://jisao.washington.edu/data/gpcp)

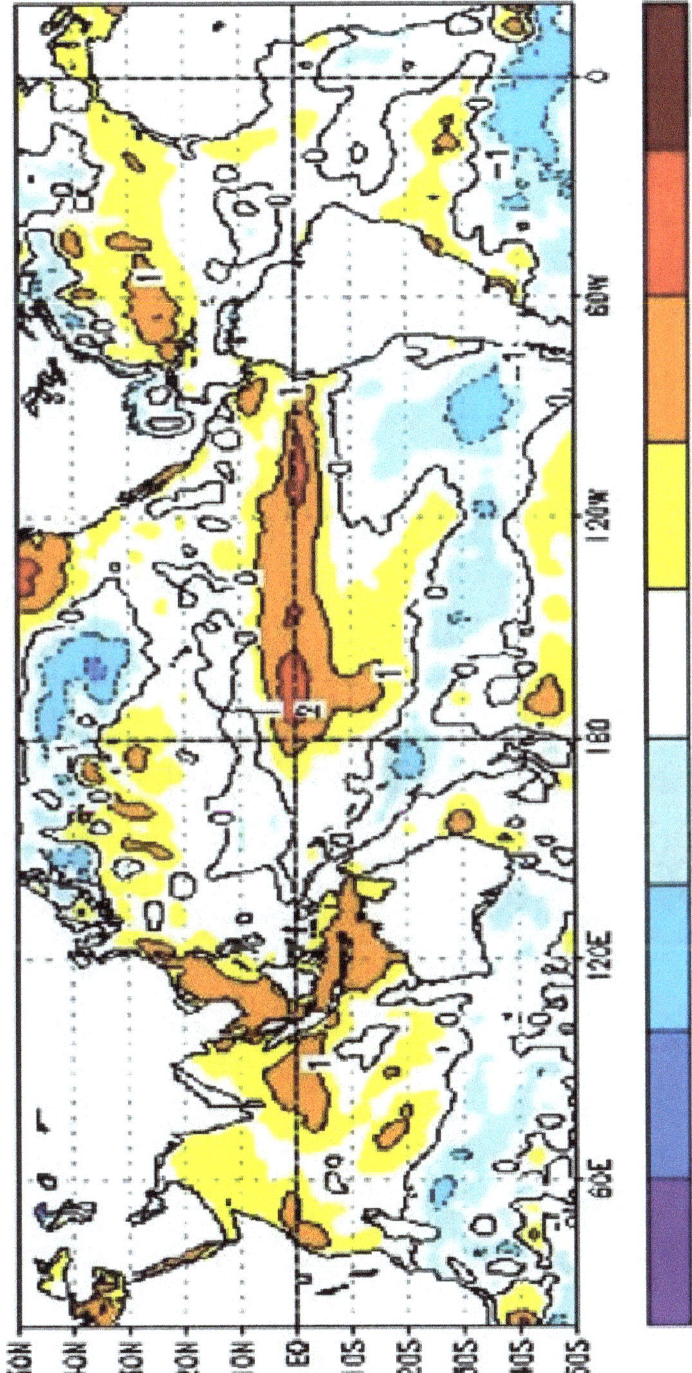

Figure 2.3.1 Positive SST anomalies (°C) over the tropical eastern Pacific Ocean in December 2002 indicative of El Nino conditions (Source: http://www.cpc.noaa.gov)

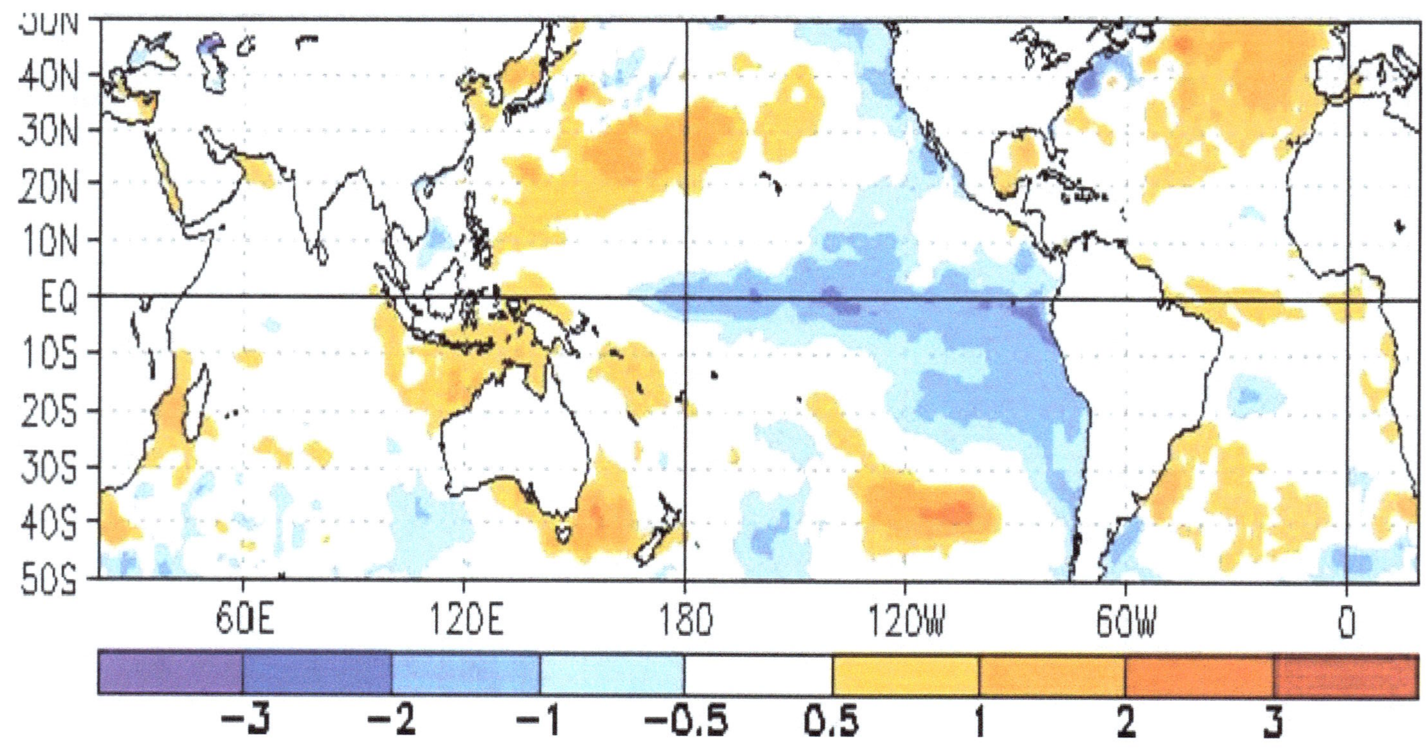

Figure 2.3.2 Negative SST anomalies (ºC) over the tropical eastern Pacific Ocean in December 2007 indicative of La Nina conditions (Source: http://www.cpc.noaa.gov)

# Chapter 2

# Global and Regional Relationships of the Monsoon

It is obvious that the Indian southwest monsoon, being such a large component of the global weather and climate system, will have positive and negative associations with other entities. Many of these associations are of a concurrent nature, having little value for monsoon prediction. The southwest monsoon may also exert a strong influence on future events in the global atmosphere or ocean. In fact, the southwest monsoon is often thought of as playing more of an active role in the climate system rather than a passive one (Kane 1997). The behaviour of the monsoon does not always resemble that of a weak slave ordered about by parameters from across the Pacific Ocean. At times the monsoon may assume the role of the master and drive the El Nino Southern Oscillation (ENSO) phenomenon instead. The monsoon and ENSO could perhaps be more aptly described as selectively interactive systems (Webster et al 1998). However, the inverse problem in which the monsoon is regarded as casting its influence on future global events, though important, has not evoked much interest among monsoon forecasters.

The origin of the term 'teleconnection' as used in the context of the monsoon, is not very certain. Gilbert Walker with whose work this term is often associated, does not appear to have used it in the several monographs and memoirs that he wrote. The credit for coining this word perhaps goes to Bjerkenes (1969).

The real monsoon teleconnections refer to those situations or developments in the land-ocean-atmosphere system which occur several months, or maybe even years, prior to the onset of the monsoon over India, and are known to exert a strong influence on the monsoon rainfall. During the time period that elapses between the detection of such signals and the arrival of the monsoon, these antecedent factors would be working together, or against each other, towards the making or unmaking of the monsoon. In a sense, the monsoon can be visualized as going through a long buildup process, during which it gets progressively manipulated by global factors that predetermine its behaviour and the amount of rain that it would produce. Thus if we know what has been 'sown' in the preceding months, we can foretell what we may 'reap' in terms of the rainfall during the four months of June to September.

## 2.1 Himalayan and Eurasian Snow Cover

In the first ever attempt to forecast the monsoon seasonal rainfall over India, Blanford (1884) used Himalayan snow cover as the sole predictor. Blanford's forecasts were based on the belief that late and heavy snowfall in the mountain regions to the north and west of India was prejudicial to the monsoon rainfall. Walker, who carried forward this work in the first quarter of the twentieth century, brought in many more predictors of the monsoon, but snow cover continued to be one of them.

In the times of Blanford and Walker, observation and estimation of snowfall could have at best been done qualitatively. Reports were received from high-altitude stations on the days when snowfall occurred. Information would be gathered about the thickness of snow in high mountain passes and about the lowest altitude up to which snowfall had reached. Walker (1909) devised a method to put together this fragmentary knowledge and to prepare a reliable snow data series for the years 1876 to 1908. He made a subtle distinction between the amount of snow that actually fell during the month of May and the amount of snow that had accumulated during the season up to the end of May. The May snowfall was categorized into no snowfall, normal, moderate, large and very large snowfall and assigned a numeric value of -1, 0, +1, +2 and +3 respectively. The snow accumulation up to the end of May was likewise categorized into very small, small, normal, moderately excess, large excess or very large excess on a 6-point scale of -2, -1, 0, +1, +2 and +3. Walker found that the correlation coefficients of these two parameters with the monsoon rainfall were -0.22 and -0.38 respectively, but the correlation coefficient between the two parameters themselves was as high as +0.53. Walker inferred that it was the increase in the snow accumulation in May rather than the snowfall in May itself that influenced the monsoon rainfall.

It was several decades later, after meteorology had entered the satellite era, that it became possible to delineate snow-covered areas of the earth with a much higher accuracy by carefully interpreting the satellite imagery (Wiesnet et al 1975). Hahn and Shukla (1976) correlated this newly available data series of satellite-derived quantitative estimates of Eurasian snow cover for the years 1966 to 1975 with the Indian summer monsoon rainfall. They found that during this period with the exception of only one year 1973, the wintertime Eurasian snow cover south of the 52° N latitude, was negatively correlated with the rainfall in the following monsoon season over India as a whole. Even on a subdivisional level, the correlations were mostly negative, and the rainfall of 10 subdivisions in the states of Gujarat, Maharashtra, Andhra Pradesh and Karnataka, which are to the south of the tracks of monsoon depressions, had the highest negative correlations with snow cover.

Subsequent to the above studies, three main long term data sets of remotely sensed snow cover have become available: (a) Nimbus-7 SMMR, (b) DMSP SSM/I and (c) EASE-Grid. The Nimbus-7 data set consists of monthly global $1° \times 1°$ snow cover and snow depth derived from the Scanning Multichannel Microwave Radiometer (SMMR) measurements over the period 1978-1987. SSMR was a 10-channel passive radiometer operating at 5 microwave frequencies of 6.63, 10.69, 18, 21 and 37 GHz, all with V and H polarizations. Since the emission from the surface at these frequencies gets modified differently by the overlying snow, the snow cover and depth can be derived indirectly. At 37 GHz, microwave radiation gets scattered by the snow in proportion to the snow depth whereas at 18 GHz, it is mainly absorbed by the snow. By using this differential response of the 18 and 37 GHz channels, the snow depth can be empirically computed from the difference in the brightness temperature of the two channels. The method works better with dry snow than wet snow and when the snow depth is more than a minimum threshold value.

The U. S. Defense Meteorological Satellite Program (DMSP) data set consists of monthly global $1° \times 1°$ and $2.5° \times 2.5°$ snow cover derived from measurements made by the Special Sensor Microwave/Imager (SSM/I) on successive DMSP satellites over the period beginning 1987 to date. SSM/I is a 7-channel passive radiometer operating at 4 microwave frequencies of 19.35, 37 and 85.5 GHz, all with V and H polarizations and 22.235 GHz with only V polarization. Snow cover scatters the microwave radiation and can be indirectly estimated from the difference in the brightness temperatures of lower (19.35 or 22.235 GHz) and higher frequency channels (37 or 85.5 GHz). Snow can be differentiated from rain by reference to the temperature of the surface. The DMSP data set provides snow cover information in terms of the fractional area of a $1° \times 1°$ or $2.5° \times 2.5°$ grid box covered by snow.

The Northern Hemisphere Equal Area SSM/I Earth Grid (EASE-Grid) Weekly Snow Cover and Sea Ice Extent data set covers the period 1978 to date. Prior to 1978, data for snow alone is available from 1966. This data set includes monthly climatology of average snow cover and sea ice, probability of occurrence and variance. The grid size is $25 \times 25$ km. Besides estimations from SSM/I and Nimbus SMMR, early snow cover values derived from a manual interpretation of visible and infrared imagery from NOAA and GOES satellites has also been incorporated in the EASE data set.

The availability of these authentic space-based data sets of snow cover and snow depth spawned many new studies to investigate afresh the relationship between Himalayan and Eurasian snowfall and the monsoon rainfall (Sankar Rao et al 1996, Kripalani et al 1996). While many of such studies had dealt with the snow cover as a single value averaged over the entire Eurasian region, Bamzai and Shukla (1999) made a finer geographical analysis on a 2° × 2° grid by using satellite-derived snow cover data for 22 years (1973-1994) and snow depth data for 9 years (1978-1987). The analysis was also done separately for northern hemispheric winter months (December to March) and spring (April and May).

Bamzai and Shukla found that large parts of Eurasia remain covered with snow in winter year after year, and that eastern and western Eurasia and the Himalayas were the only regions that exhibit interannual variability of winter snow cover. After the decrease in snow cover from winter to spring, the Himalayan and eastern Eurasian regions continue to show high interannual variability. For the 1973-1994 period, a significant inverse correlation of -0.63 was found between anomalies of monsoon rainfall and winter snow cover over a large coherent region of western Eurasia bounded by 40-60°N, 10°W-30°E, that persisted into the spring season. The correlation decreased to -0.34 if a larger Eurasian region of 20-90°N, 0-190°E was considered, and further down to -0.20 for southern Eurasia, covering 20-50°N, 0-190°E. There were no regions of significant positive correlation, but the winter snow cover in the Himalayan region was found to have a small positive correlation of +0.20, quite contrary to what had been traditionally thought of. The correlation coefficients between Eurasian snow cover anomalies and the Indian monsoon rainfall anomalies averaged separately for the months of December to May, December to March, and April and May, were -0.39, -0.34 and -0.28, respectively.

That the monsoon rainfall is related differently to different regions of the northern hemisphere was also seen by Kripalani et al (1999) in their study of Soviet snow depth data sets for 1881-1995. They identified two coherent regions, one in western Eurasia and another in eastern Eurasia, where the snowfall showed a negative and positive relationship respectively with the monsoon rainfall, the strongest relationships being seen in January.

Kripalani et al (2003a) reexamined the monsoon-snow relationship using INSAT-derived snow estimates over the western Himalayas for the period 1986-2000. They found that the spring (February to May) snow cover (snow melt) area is negatively (positively) related with subsequent summer monsoon rainfall over India. This implies that a smaller snow cover area and faster snow melt is conducive for good monsoon activity over India. The impact of snow melt on monsoon rainfall and circulation appears to be stronger than that of snow cover. Kripalani et al found that the negative relationship of the monsoon rainfall with the snowfall of the preceding winter seemed to have been broken in recent times and changed sign to a positive relationship. However, the negative relationship between the monsoon rainfall and the preceding spring snowfall still holds good.

As satellite-based snow data sets get better in the coming years, both in terms of quality and quantity, it is only to be expected that more detailed statistical studies of the above type would become available. However, as is the case with many other global relationships of the monsoon, refined statistics are not by themselves able to throw a better light on the underlying physical processes of the land, atmosphere and ocean. Correlation figures cannot tell us why the monsoon rainfall is influenced by the snowfall over only certain preferred parts of the vast Eurasian continent or in certain preferred months. What is surprising, however, is that after Blanford had used Himalayan snow cover to predict the monsoon rainfall, a whole century had to elapse before a plausible explanation of the physical process was forthcoming.

We now know that the earth's snow cover dominates atmospheric and soil processes in several different ways. Because of its high albedo, snow reflects a very large fraction of the incoming solar radiation and does not allow it to enter the soil. As it has a low heat conductivity, it serves as a thermal insulation between the atmosphere and the land surface. When the snow melts, it acts both as a heat sink and as a source of fresh water. However, the various links in the chain between Eurasian snow cover and monsoon rainfall are yet to be understood.

Shukla (1987) hypothesized that an excessive snowfall would lead to a delay in the buildup of the monsoon temperature gradient, because a larger part of the solar energy would be reflected and a smaller part left for warming the surface. Vernekar, Zhou and Shukla (1995) carried the investigation further with the aid of the COLA GCM and the Nimbus-7 SSMR-derived snow depth data. The model was integrated in an annual mode from February, when the Eurasian snow cover is at its maximum extent, up to September, when the southwest monsoon season comes to a close. Two experiments were made, by initially prescribing a heavy snow cover and a light snow cover, with the model predicting snow accumulation and snow melting during integration. The results showed that excessive Eurasian snow cover causes a reduction of the midtropospheric meridional temperature gradient across 0-30° N which leads to a weakening of the monsoon circulation, delay in the monsoon onset as well as its withdrawal and decrease in the rainfall. However, the effects are not confined to the Indian region alone, as there is a general weakening of the Somali jet, lower tropospheric westerlies and the trade winds in the equatorial eastern Pacific Ocean.

Robock et al (2003) re-examined the relationship between interannual variations of the strength of the monsoon with not only snow cover but other surface conditions over Eurasia such as soil moisture, surface air temperature and atmospheric circulation features. Using a precipitation data set spanning 1870-2000, they found that for the periods 1870-1895 and 1950-1995, strong Indian summer monsoon precipitation was preceded by warmer than normal temperatures over Europe and North America in the previous winter and over western Asia in the previous spring, but by colder temperatures over Tibet. They confirmed that heavy monsoon rainfall is preceded by lower than normal snow cover in the Eurasian region but by higher than normal snow cover in the Tibetan Plateau during the period of their analysis. The snow cover anomalies thus reflect a wave-like pattern, suggesting that they influence the monsoon rainfall through circulation anomalies rather than directly. Reebock et al also found that the soil moisture is not strongly related to the strength of the monsoon. The soil moisture pattern does not match the snow pattern, nor does it show a response to anomalous snow cover. They reasoned that while the snow-albedo feedback is always operating, the snow by itself does not physically control the monsoon.

Dash et al (2005, 2006) used a spectral GCM to conduct sensitivity experiments with two contrasting Eurasian snow depths. They showed that because of the west Eurasian snow anomalies, the midlatitude circulations in winter and spring show significant changes in the upper and lower level wind, geopotential height, velocity potential and stream function fields. Such changes in the large scale circulation pattern may be interpreted as precursors to weak (strong) monsoon circulation and deficient (excess) monsoon rainfall.

In summary, we do not yet have a clear understanding of the physical processes that link the monsoon rainfall to Eurasian snow cover. Neither are statistical relationships are of any greater help, as depending upon the specific Eurasian region we choose, we may find the correlations to be inverse or direct, weak or strong, present or absent. No wonder then, that snow-related predictors which were the favourites of Blanford and Walker, were given up by IMD in the middle of the twentieth century, then reinstated in 1988, only to get dropped again from the revised IMD models of 2003 and 2007.

## 2.2 Southern Oscillation

The extensive work of Walker (1923, 1924) in search of correlations of seasonal variations in world weather, led him to the discovery of three great pressure oscillations which he named as the north Atlantic, north Pacific, and southern oscillations and described as follows:

The north Atlantic oscillation was an accentuation of the pressure difference between the Azores and Iceland in autumn and winter and an associated strong circulation of the winds in the Atlantic, a strong Gulf Stream, high temperatures in winter and spring in Scandinavia and the east coast of the United States, and with lower temperatures on the east coast of Canada and the west of Greenland.

The north Pacific oscillation at first sight seemed to bear a close resemblance to the north Atlantic oscillation, but the meteorological observations were either wholly lacking or fragmentary and it was therefore impossible to trace the resemblance further than that of a very general nature.

The southern oscillation implied the tendency of pressure at stations in the Pacific (San Francisco, Tokyo, Honolulu, Samoa and South America) and of rainfall in India and Java to increase, while pressure in the region of the Indian Ocean (Cairo. northwest India, Port Darwin, Mauritius, southeast Australia and the Cape) decreased.

Many decades later, Bjerknes (1969) explained that while the Indian Ocean had no equatorial cold water as such, the Pacific equatorial cold water belt was the largest and coldest of its kind. Therefore, when the cold water belt along the equator is well-developed, the air above it is too cold and heavy to join the ascending motion in the Hadley circulations. Instead, the equatorial air flows westward between the Hadley circulations of the two hemispheres to the warmer western Pacific. Bjerkenes assumed that the gradient of sea temperature along the equator is the cause of this east-west circulation which he named as the 'Walker circulation' as it can be shown to be an important part of the mechanism of Walker's southern oscillation.

The Walker circulation has an ascending branch over the warm waters of the western Pacific near 165° E and a descending branch over the cold waters of the eastern Pacific around 90° W (Figure 2.2.1). However, when the longitudinal temperature gradient and the related pressure gradient are disrupted by the cessation of equatorial upwelling, the axis of the Walker circulation shifts eastward. This allows the ascending branch to produce cloudiness and rain over normally suppressed regions (Figures 2.2.2 and 2.2.3). Bjerkenes thus integrated the Walker circulation and the southern oscillation into one overall mechanism that operated over the Pacific and Indian Ocean regions and showed itself in the distribution of pressure, temperature and rainfall over that region.

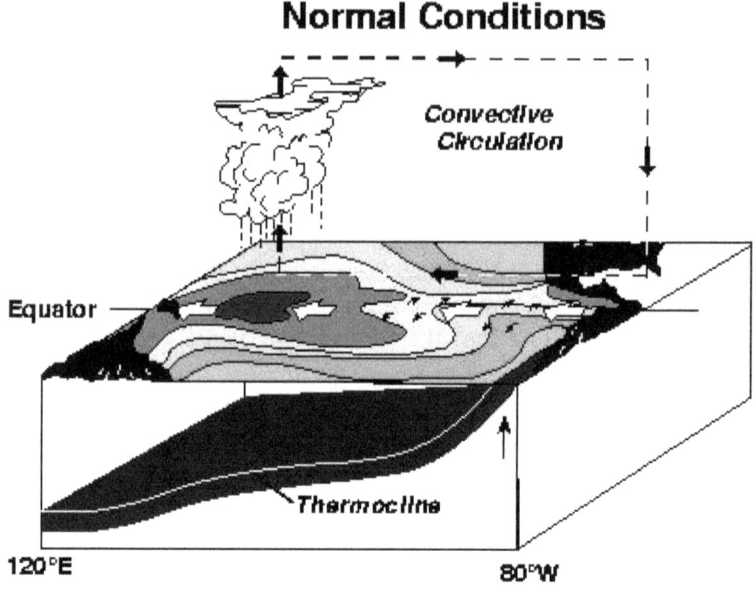


Figure 2.2.1 The east-west Walker circulation under normal conditions
(Source: http://www.pmel.noaa.gov)

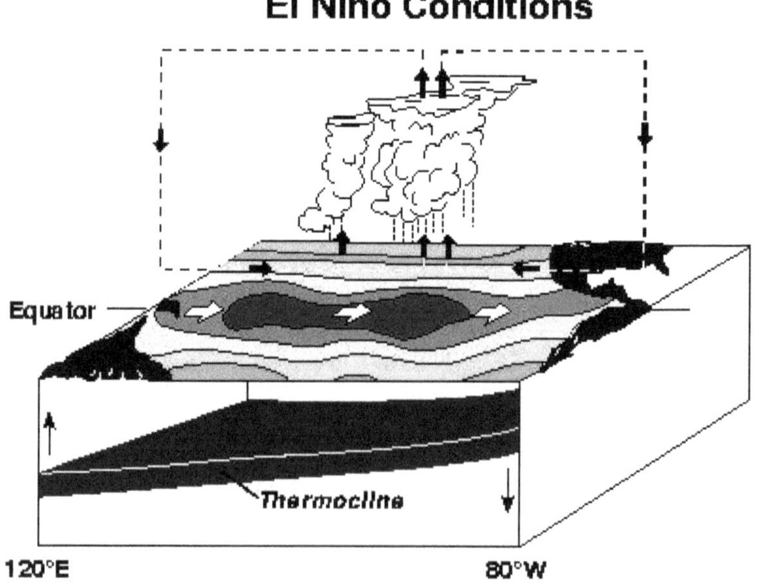


Figure 2.2.2 The east-west Walker circulation under El Nino conditions
(Source: http://www.pmel.noaa.gov)

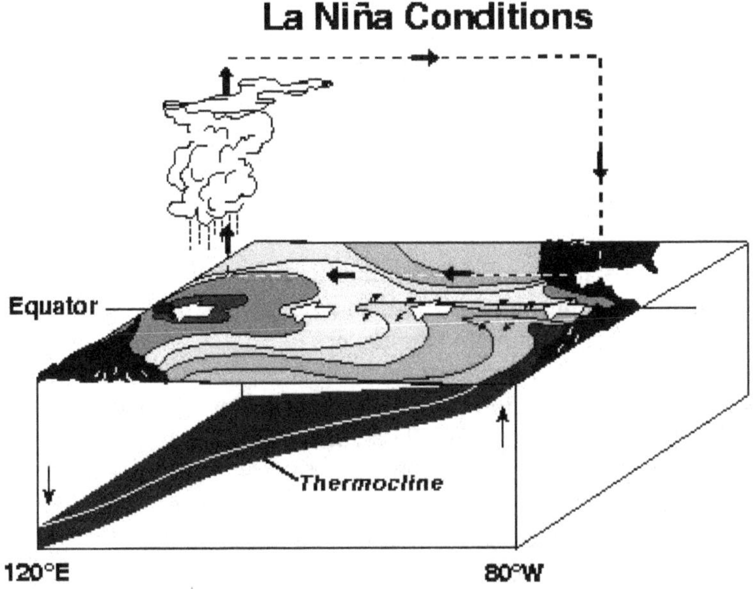


Figure 2.2.3 The east-west Walker circulation under La Nina conditions
(Source: http://www.pmel.noaa.gov)

Shukla et al (1983) reasoned that most of the predictors that Walker used in his statistical models for monsoon forecasting were in fact just manifestations of different aspects of the southern oscillation. Analysing an 81-year data series, they found that the monsoon seasonal rainfall was very well-correlated to the trend of the Darwin pressure anomaly before the monsoon season, though not to the anomaly itself. If the trend from the months December-February to March-May is negative, the monsoon rainfall is heavy, and if this trend is positive, the monsoon rainfall is deficient.

It is now an accepted practice to describe the strength and phase of the southern oscillation by the Southern Oscillation Index (SOI) which is defined as the standardized anomaly of the difference in surface pressure reported at two stations, Tahiti in the Pacific and Darwin in Australia. SOI is usually calculated on a monthwise basis as follows:

$$SOI_{month} = 10\ [\Delta P - \Delta P_{ave}]\ /\ [SD\ (\Delta P)]$$

where $\Delta P$ is the difference between average mean sea level pressure at Tahiti and that at Darwin for a given month, $\Delta P_{ave}$ is the long term average of $\Delta P$ for that month, and $SD(\Delta P)$ is the long term standard deviation of $\Delta P$ for that month. The multiplying factor 10, which is optional, makes it possible for the SOI to be expressed as an integer value ranging from -35 and +35. When Tahiti has a negative pressure anomaly and Darwin has a positive pressure anomaly, the southern oscillation will be in its negative phase, and when the reverse happens, it will be in its positive phase.


## 2.3 El Nino, La Nina and ENSO

Sea surface temperatures off the coastlines of Peru and Ecuador in South America exhibit an annual cycle, with an average maximum temperature of 19 °C in February and minimum temperature of 15.5 °C in October (Deser et al 1987). An ocean current develops in this region in December around Christmas time and it is known locally in Spanish as El Nino, meaning the baby boy Jesus. In some years, the Peru current extends further southward to 12° S, when it can destroy plankton and fish populations and cause the fisheries business to suffer losses. Such instances which are characterized by positive sea surface temperature anomalies along the Peru-Ecuador coast have come to be known as El Nino events in meteorological and oceanographic parlance. Situations of an opposite nature wherein the sea surface temperature anomalies are negative, meaning a colder than normal ocean, are called La Nina, the Spanish word for a baby girl.

These seemingly innocuous local events that occur in the eastern Pacific Ocean, however, set into motion a chain of atmospheric and oceanic ramifications that can be far-reaching in terms of both space and time. Though the influence of El Nino on global weather and climate had been known for long, the fact that it can produce major upsets in the normal atmospheric and oceanic processes was brought into limelight by Rasmusson et al (1982, 1983). They analyzed the monsoon rainfall data over India and Sri Lanka over the 105-year period 1875-1979, during which they identified a total of 25 strong and moderate El Nino events, averaging one every 4.2 years. They found a strong tendency for the Indian monsoon rainfall to be below normal in the El Nino years, with 20% of the El Nino years accompanied by catastrophic droughts and 44% of the El Nino years by moderate droughts, implying of course that not all El Nino events necessarily produce a monsoon drought. It is now known that the temperature-related El Nino/La Nina episodes go hand in hand with the pressure 'see-saw' that constitutes the southern oscillation (Figures 2.3.1 and 2.3.2). During an El Nino, the southern oscillation is in a negative phase, when positive pressure anomalies occur over Indonesia and the western part of the tropical Pacific Ocean while there are negative pressure anomalies over the eastern part of the tropical Pacific Ocean. The reverse situation is observed in a La Nina episode. When a prolonged spell of negative SOI coincides with the existence of an El Nino, the situation is described as an ENSO (El Nino Southern Oscillation) event (Figure 2.3.3).

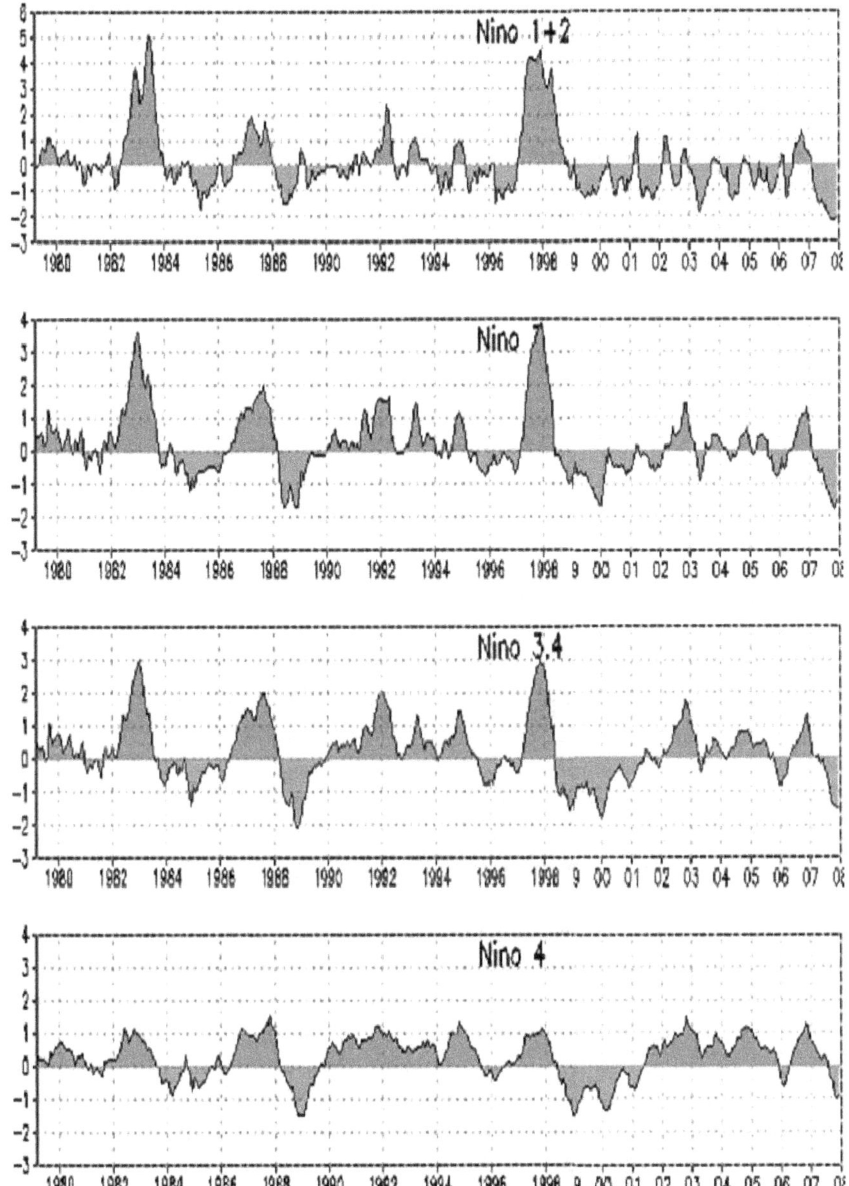

Figure 2.3.3 SST anomalies (°C) over the four El Nino regions
Nino 1+2, 3, 3.4 and 4 from 1979 to 2008 (Source: http://www.cpc.noaa.gov)

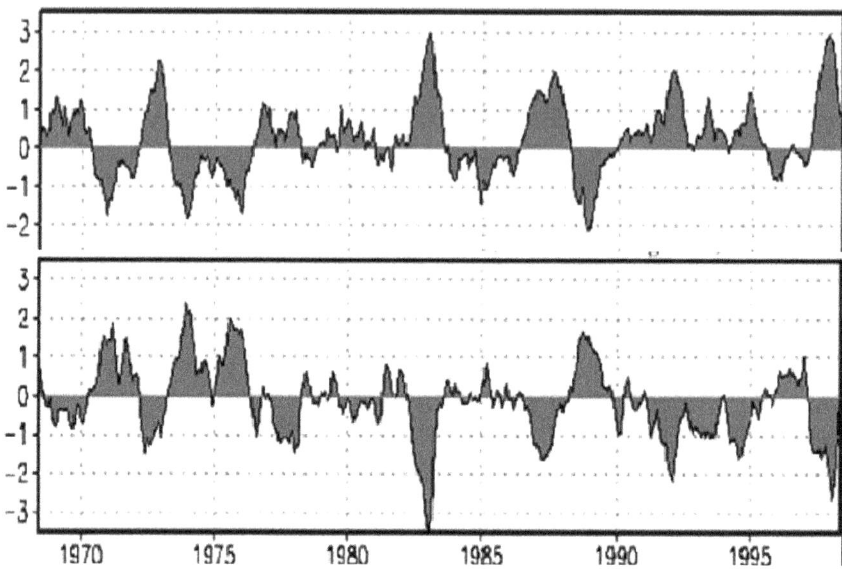

Figure 2.3.4 The ENSO cycle showing the inverse relationship between El Nino/La Nina and the phase of the southern oscillation. Graph above shows SST anomalies (°C) for the Nino 3.4 region. Graph below shows the 3-month running mean of the SOI. (Source: http://www.cpc.noaa.gov)

The El Nino and La Nina episodes typically have a 12-month life cycle that begins in June and ends next May with a peak around December. The timing of this cycle has an important bearing on the long range forecasts of the Indian southwest monsoon. Such forecasts have to be issued in May, when the El Nino/La Nina episodes have not commenced. Hence the monsoon forecasts have to factor in the likely evolution of the El Nino or La Nina concurrently with the ensuing monsoon. What complicates the problem further is that the El Nino/La Nina episodes vary widely in their characteristics like time of commencement, region of occurrence, peak strength, timing of the peak, and total lifetime, each of which has a different influence on the monsoon. El Nino and La Nina do not alternate every year but have a quasi-periodicity of 3 to 7 years. The events are categorized as weak, moderate or strong depending upon the SST anomalies being less than 1 °C, between 1 and 2 °C, or larger than 3 °C, with each category having a different nature of global influences. Some prolonged episodes have persisted for more than two years at a stretch, and in many years conditions have remained neutral with near-zero SST anomalies.

Like El Nino and La Nina, the ENSO cycle is also quasi-periodic, varying between 2 and 7 years, and having an average periodicity of about four years, and there is large variability in the occurrence of ENSO events in the recent past. Between 1981 and 2000, there were 5 El Nino episodes (1982-83, 1986-87, 1991-93, 1994-95 and 1997-98) and 3 La Nina episodes (1984-85, 1988-89 and 1995-96). The El Nino events of 1982-83 and 1997-98 were the strongest among them. The El Nino episodes of 1991-93 and 1994-95 occurred in succession without the development of an intervening La Nina (Figure 2.3.3).

In 1997-98, soon after the publication of the work of Rasmusson et al (1983) correlating droughts over India to El Nino, came the development of an extremely strong El Nino and equally strong fears began to be expressed that there would be a major drought over India. However, this did not happen, and in 1998, India received above-normal rainfall instead. Elsewhere too, the impacts were not quite as expected. In a critical analysis of the 1997-98 El Nino event, Kane (1997) adopted finer definitions of ENSO with the aim of unravelling the reasons as to why every ENSO event has a somewhat different behaviour and why every El Nino does not lead to a monsoon failure over India.

Starting from 1871, Kane categorized the ENSO events into three broad categories: (1) Unambiguous El Nino (ENSOW-U) in which the El Nino existed near the Peru coast, and the SOI minimum and the Pacific SST maximum occurred during May-August, (2) Ambiguous El Nino (ENSOW-A) in which the El Nino existed near the Peru coast, but the SOI minimum and the SST maximum occurred in the early or later part of the calendar year, not in the middle, and (3) Other El Nino (EN) in which the El Nino existed near the Peru coast, the SOI was minimum, but SST was neither warmer nor colder, but just normal. Kane also categorised the remaining years into several other categories like EN but no SO, SO but no EN, La Nina and so on. Kane concluded that the ENSOW-U events have an almost 1-to-1 association with droughts over India, while the ENSOW-A events are mostly associated with normal rainfall. Another interesting conclusion drawn by him was that the strength of the El Nino was not of much relevance to the Indian monsoon rainfall. All this only goes to confirm that ENSO is not the only parameter that drives the Indian monsoon and there are other factors that can offset or counterbalance its effects.

While the phenomena of El Nino and La Nina are undoubtedly of great importance to monsoon forecasting, the main difficulty lies in anticipating how they may behave concurrently with the monsoon. Thus good El Nino prediction becomes a prerequisite to successful monsoon forecasting. While a lot of effort is under way in this direction, ENSO forecasts generated by different models show considerable spread. There is therefore a need to reduce uncertainties in ENSO forecasts so that they can be used with more confidence (McPhaden 2004).

## 2.4 Land Surface Temperature

The monsoon having been traditionally looked upon as a thermally driven large scale circulation, attempts have always been made to examine its relationships with temperatures over the land and sea surface. Mooley et al (1988) had found that the mean monthly minimum temperature over northwest India in May had a significant positive correlation with the monsoon rainfall. Krishna Kumar et al (1997) had carried out a similar study on the premonsoon thermal field on a wider spatial scale.

Kripalani et al (2003b) used the anomalies of the January-February average of Northern Hemisphere Surface Temperature (NHST) as an indicator of global warming. The relationship between the Indian monsoon rainfall and NHST has a positive sign. An increase in NHST may intensify the land-ocean temperature contrast and the meridional Hadley circulation and favour monsoon activity over India. It was found that the 11-year sliding correlation during the period 1960-1980 was strong, after 1980 it started weakening, around 1990 it became zero and changed sign after that. If global warming had really been the cause of the recent ENSO-monsoon weakening, then the relationship between NHST and monsoon rainfall should have strengthened in recent times. The findings of Kripalani et al do not support global warming as a cause of the ENSO-monsoon weakening.

Liang et al (2005) have examined the role of land-sea distribution in the formation of the Asian summer monsoon through a series of interesting and idealized numerical experiments. Their results show that the existence and geometric shape of land-sea distribution crucially affects the Asian summer monsoon. In the hypothetical situation of there being only the subtropical Eurasian landmass and no ocean, a weak summer monsoon may develop over its southeastern corner, but there would be no tropical summer monsoon. In the opposite case of an 'aquaplanet', no monsoon would develop at all. It is the existence of tropical land that induces cross-equatorial flow and strong low level southwesterlies over the tropical regions, leading to the formation of the Asian summer monsoon over India, the Bay of Bengal and the South China Sea.

## 2.5 Sea Surface Temperature

Out of the many global relationships of the monsoon, the ENSO-monsoon relationship is one that has been the most investigated, and naturally most favoured for use as a predictor. Many statistical models for long range forecast of monsoon rainfall look for the change in magnitude of ENSO indices from winter to spring and incorporate it in their computations. However, the response or rather the lack of response of the monsoon to the major El Nino event of 1997-98 set in motion many investigations into what was interpreted as a weakening of the ENSO-monsoon relationship. If it was indeed weakening, then there surely must be signs of the monsoon developing closer associations elsewhere, such as the western and northwest Pacific Ocean, the Indian Ocean or the north Atlantic Ocean. This possibility led to a spurt of studies of the sea surface temperatures over these regions to discover afresh their relationships with the monsoon that had been hitherto neglected.

Sahai et al (2003) looked at this problem in a most generalized manner by correlating the monsoon rainfall with sea surface temperatures over a global grid. First, 107 oceanic regions of 5° latitude × 10° longitude size were identified for their significant and consistent correlations with the monsoon rainfall with lags of different time periods in terms of 3-month seasons. After a screening for various statistical tests of significance, 14 of these were ultimately selected as predictors, 1 from the Indian Ocean, 8 from Pacific and 5 from Atlantic. Significantly, there were no predictors from the north Indian Ocean and the east Pacific in this final set. The predictors had lag periods ranging from a year to four years, so that a monsoon forecast can possibly be made as much as nine months in advance.

The work of Sahai et al is significant in that it offers statistical evidence to confirm the weakening of the monsoon-ENSO relationship and to bring into focus the existence of much stronger relationships of the monsoon with other ocean basins. They have stressed that their 14 SST-based predictors have the ability to capture the important heat sources and sinks in the coupled ocean-atmosphere system of the tropics and extratropics, and the biennial oscillation of the coupled system modulated in turn by longer term oscillations. However, the analysis is purely empirical and no 'cause-and-effect' explanation has been offered about how the 14 global SST predictors are able to influence the monsoon rainfall, some of them after two, three or even four years.

In a later but very similar effort, Pai et al (2005) also showed that the monsoon rainfall had strong correlations with the SST over regions of the Indian, Pacific and Atlantic Oceans and they could isolate 13 SST-based predictors with lags varying from a season to four years. These results are also empirical and have quite the same advantages and drawbacks as of the earlier work of Sahai et al.

There have been other studies of monsoon rainfall in relation to SST outside the ENSO-dominated region of the Pacific. For example, Reddy and Salvekar (2003) studied the eastern equatorial Indian Ocean, where anticyclonic twin gyres evolve in the month of May and migrate towards the western part of the Indian Ocean. They found a negative correlation between June SST over this region and the July-September rainfall over India. In another attempt to find a predictor in SST, Rajeevan et al (2002) have noticed a positive correlation between the monsoon rainfall and SST anomalies over the Arabian Sea in winter and again in May, and over southeast Indian Ocean during February-March.

## 2.6 Location of the 500 hPa Ridge

The east-west alignment of the 500 hPa ridge over India and its neighbourhood separates the westerly wind regime to its north from the easterly tropical regime to the south (Figure 2.6.1). The normal alignment of the 500 hPa ridge over this region is approximately along 12° N in January, moving northward to 15° N in April and 28° N in July and returning to 19° N in October.

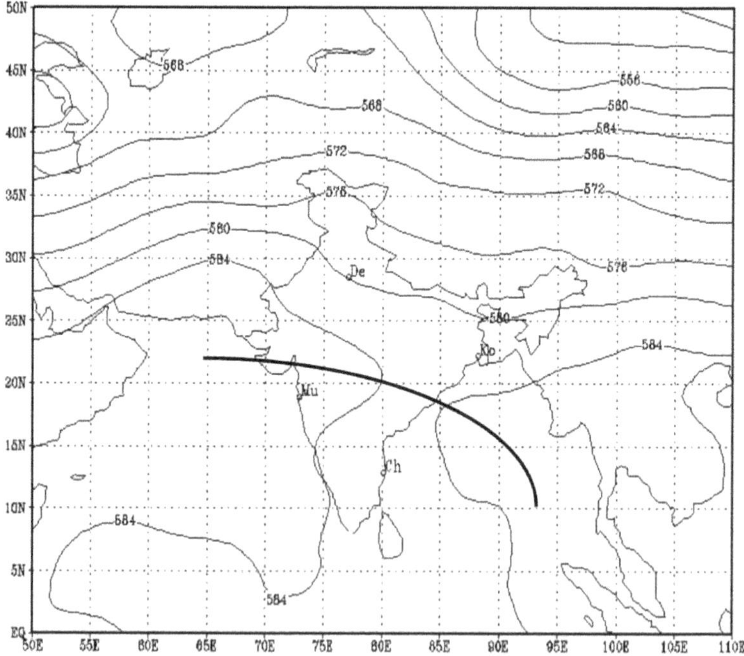

Figure 2.6.1 500 hPa contour analysis for 10 April 2008 1200 UTC and the 500 hPa ridge (Source: http://www.imd.gov.in)

The possibility of a connection between the location of the 500 hPa ridge in April over India and the strength of the subsequent monsoon was first brought out by Banerjee et al (1978). In a later and detailed corroboration of this result, Mooley et al (1986) found a significant positive correlation between the all-India monsoon rainfall and the location of the 500 hPa ridge along the 75° E longitude averaged for the month of April. The normal position of the ridge over 1939-80 worked out to be at 15.4° N latitude. However, in years of deficient monsoon rainfall, its position was found to have shifted 3° to the south of this normal position, and in excess rainfall years, it was 1° to the north instead.

Mooley et al (1986) interpreted the position of the ridge as a measure of the influence of the westerly troughs on the upper tropospheric thermal conditions over central and north India. A southward shift of the ridge position signifies the persistence of colder than normal conditions which can affect the monsoon adversely. In a continuation of this study, Shukla and Mooley (1987) confirmed this hypothesis and derived a regression equation between the monsoon rainfall and two parameters, one of which was the location of the 500 hPa ridge in April and the other was the Darwin pressure tendency from January to April.

## 2.7 Sunspot Activity

Ever since the discovery of the 11-year sunspot cycle, whether and how it modulates the earth's weather and climate has been a matter of interest and investigation. In the past, three extended periods are known to have occurred during which sunspots were very few in number or totally absent, and these have been named as the Sporer minimum (1450-1540), the Maunder minimum (1645-1715) and the Dalton minimum (1795-1820). These three periods precisely match the time when the earth is said to have experienced the Little Ice Age. Since the sun is undoubtedly the driving force behind the earth's climate system, such correlations must exist on the interglacial time scale, but what happens on shorter time scales is not so evident.

One particular object of the studies of the solar-terrestrial climate relationship is that if such a link can be established, then an attempt can be made to forecast atmospheric conditions several years ahead on the basis of likely variations in the sunspot activity. This, however, requires that we develop a cause-and-effect understanding of the process through which sunspots, and changes in their number, exert their influence on the earth.

In recent years, research in this field has gathered fresh momentum on two counts. The first is the launch of satellites with the exclusive purpose of monitoring the sun, resulting in the availability of detailed, precise and continuous data sets on solar irradiance and sunspot activity. The Active Cavity Radiometer Irradiance Monitor (ACRIM-I) was flown in 1980 and worked until 1989 as the Solar Maximum Mission. Its successor satellite ACRIM-III, launched in December 1999 carried an instrument to measure the total solar irradiance. This satellite provided the first data set that clearly demonstrated that the total radiant energy from the sun was not a constant. The Solar Radiation and Climate Experiment (SORCE) is a NASA-sponsored satellite mission that provides state-of-the-art measurements of incoming x-ray, ultraviolet, visible, near-infrared, and total solar radiation. The measurements provided by SORCE specifically address long term climate change, natural variability and enhanced climate prediction, and atmospheric ozone and UV-B radiation. SORCE was launched by NASA in 2003 into a 645 km, 40° inclination orbit and carried four different instruments.

The second reason is the need to know whether sunspots could be a natural cause behind the currently observed warming trend rather than human-induced increase in greenhouse gases. As a result, there is a spurt of new research on various aspects of the problem. Attempts have been made for example, to establish correlations with the length of the solar cycle instead of sunspot numbers (Friis-Christensen et al 1991), to study the variations in solar luminosity (Foukal et al 2006), to examine the North Atlantic Oscillation, QBO and ENSO relative to the solar cycles (Moore et al 2006), and to connect cosmic rays with cloud cover (Carslaw et al 2002). Empirical studies suggest that the sun could not have contributed more than 30% to the recently observed warming (Solanki et al 2003).

Walker (1923) made an extensive analysis of the correlation between the annual sunspot number and the meteorological observations at stations around the globe. He also used the prominence data recorded at the Kodaikanal Observatory. Walker surmised that if high sunspot numbers meant an increased direct heating of the earth, it would lead to increased evaporation over the ocean and higher rainfall over land, while its indirect effect would be seen in the form of forced pressure oscillations. However, having derived an array of linear correlations with diverse values and signs, Walker could not have arrived at a very consistent conclusion. He found that high sunspot numbers were associated with a general decrease in annual temperature, particularly in the tropics, but they were associated with an increase as well as decrease in annual pressure and rainfall over different regions.

It was many decades later that there was a revival of interest in the association of sunspot activity with the monsoon rainfall (Jagannathan and

Bhalme 1973). They studied the pattern of rainfall distribution during the 25 pentads forming the monsoon season (31 May to 2 October) for each of the years 1901-51 for a network of 105 stations over India. The variations of the pentad rainfall during the monsoon season were represented analytically as a function of time in terms of a set of six 'distribution parameters' which represented the mean rainfall, the linear trend in the rainfall with the advance of the season, and other more complex features of the rainfall distribution. Of the several periodicities exhibited in the individual spectra, the QBO was predominant but the cycles corresponding to the sunspot cycle were also there. Since sunspot cycles have been observed in several other atmospheric phenomena, their existence in monsoon rainfall distribution was considered as probable even though the exact cause-effect relationship was obscure.

Jagannathan et al found that the mean rainfall was larger during sunspot maximum than during sunspot minimum over north Assam, north Bengal, Bihar, sub-Himalayan east U.P. and the central parts of the peninsula. Over the rest of the country, the rainfall during the sunspot minimum was larger than that during the sunspot maximum. In addition, oscillations in the mean rainfall corresponding to the sunspot cycle were dominant in the areas where the rainfall was influenced by orography. The rainfall distribution patterns during the monsoon season varied from one sunspot epoch to another. A significant result of the study was that during a sunspot maximum the core of the negative pressure departures was near the foot of the Himalayas, while during a sunspot minimum, this core was situated over central India, indicative of a possible correlation with the active/break cycle of the monsoon. It was suggested that the response of the rainfall to the sunspot epochs could be through the changes possibly induced in the atmospheric circulation. However, the mechanism through which the energetically minute changes in the solar activity could bring about the much larger changes observed in the atmosphere was a question that was left unaddressed.

Bhalme et al (1981, 1984) derived a family of Drought Area Indices (DAI) and Flood Area Indices (FAI) for India based upon monthly monsoon rainfall data for 1891-1979. These indices were subjected to a power spectrum analysis, which revealed a highly significant 22-year cycle in the FAI, that was in phase with the 22-year double (Hale) sunspot cycle. The double cycle takes into account the alternating polarity of the leading sunspots and the sign of the spot magnetic field in the 11-year cycle. Bhalme et al found that all large scale flood events over India, resulting from intense monsoon activity, occurred in the major sunspot cycle and none in the minor sunspot cycle. Significantly, large scale drought events over India did not show such a preference to the phase of the sunspot cycle.

## 2.8 The Predictive Value of Teleconnections

From the times of Blanford and Walker until now, monsoon forecasters have been searching far and wide, on different spatial as well as temporal scales, for reliable monsoon teleconnections that could lead to accurate monsoon prediction. The teleconnections are essentially of three types: (a) physically explainable processes, (b) statistically significant relationships, and (c) known physical processes which are also statistically well-correlated. Teleconnections of type 'a' certainly improve our understanding of the monsoon phenomenon, and may also help us to forecast the rainfall in a qualitative sense. The type 'b' teleconnections can be converted into a statistical regression model that can yield a quantitative forecast of monsoon rainfall, without necessarily getting a physical insight into the process. The type 'c' teleconnections are the most ideal ones and there has been a continuing search for this category of predictors or parameters as they are often called. Unfortunately, for reasons not fully understood, the statistical regressions degrade with the passage of time, and parameters which were once considered ideal for prediction purposes have to be abandoned later. This is the reason why statistical models for long range forecasting of monsoon seasonal rainfall have not reached the desirable level of accuracy, their skill scores have not improved, and they appear to be in need of constant refinement. This aspect will be discussed later in Chapters 3 and 6.


## 2.9 References


Bamzai A. S. and Shukla J., 1999, "Relation between Eurasian snow cover, snow depth, and the Indian summer monsoon: An observational study", *J. Climate*, 12, 3117-3132.

Banerjee A. K., Sen P. N. and Raman C. R. V., 1978, "On foreshadowing southwest monsoon rainfall over India with midtropospheric circulation anomaly of April", *Indian J. Meteorology Geophysics*, 29, 425-431.

Bhalme H. N. and Jadhav S. K., 1984, "The double (Hale) sunspot cycle and floods and droughts in India", *Weather*, 39, 112-116.

Bhalme H. N. and Mooley D. A., 1981, "Cyclic fluctuations in the flood area and relationship with the double (Hale) sunspot cycle", *J. Applied Meteorology*, 20, 1041-1048.

Bjerknes, J., 1969, "Atmospheric teleconnections from the equatorial Pacific", *Monthly Weather Review*, 97, 163-172.

Blanford H. F., 1884, "On the connection of the Himalayan snow with dry winds and seasons of droughts in India", *Proc. Royal Society*, 37, 3-22.

Carslaw K. S., Harrison R. G. and Kirkby J., 2002, "Cosmic rays, clouds and climate", *Science,* 298, 1732-1737

Dash S. K., Parth Sarthi P. and Panda S. K., 2006, "A study on the effect of Eurasian snow on the summer monsoon circulation and rainfall using a spectral GCM", *Int. J. Climatology*, 26, 1017-1025.

Dash S. K., Singh G. P., Shekhar M. S., Vernekar A. D., 2005, "Response of the Indian summer monsoon circulation and rainfall to seasonal snow depth anomaly over Eurasia", *Climate Dynamics*, 24, 1-10.

Deser C. and Wallace J. M.,1987, "El Nino events and their relation to the Southern Oscillation: 1925-1986", *J. Geophysical Research,* 92, 14189-14196.

Foukal P., Fro¨hlich C., Spruit H. and Wigley T. M. L., 2006, "Variations in solar luminosity and their effect on the Earth's climate", *Nature*, 443, 161-166.

Friis-Christensen E. and Lassen K., 1991, "Length of the solar cycle, an indication of solar activity closely associated with climate". *Science*, 254, 698-700.

Hahn D. G. and Shukla J., 1976, "An apparent relationship between Eurasian snow cover and Indian monsoon rainfall", *J. Atmospheric Science*, 33, 2461-62.

IMS, 1986, *Sir Gilbert Walker Selected Papers, Long Range Forecasting of Monsoon Rainfall*, Indian Meteorological Society, New Delhi, 254 pp.

Jagannathan P. and Bhalme H. N., 1973, "Changes in the pattern of distribution of southwest monsoon rainfall over India associated with sunspots", *Monthly Weather Review*, 101, 691-700.

Kane R. P., 1997, "Relationship of El Nino/Southern Oscillation and Pacific sea surface temperature with rainfall in various regions of the globe", *Monthly Weather Review*, 125, 1792-1800.

Kripalani R. H., Singh S. V., Vernekar A. D. and Thapliyal V., 1996, "Empirical study on Nimbus-7 snow mass and Indian summer monsoon rainfall". *Int. J. Climatology,* 16, 23-34.

Kripalani R. H. and Kulkarni A., 1999, "Climatology and variability of historical Soviet snow depth data: Some new perspective in snow-Indian monsoon teleconnections", *Climate Dynamics*, 15, 475-489.

Kripalani R. H., Kulkarni A. and Sabade S. S., 2003a, "Western Himalayan snow cover and Indian monsoon rainfall: A re-examination with INSAT and NCEP/NCAR data", *Theoretical Applied Climatology*, 74, 1-18.

Kripalani R. H., Kulkarni A., Sabade S. S. and Khandekar M. L., 2003b, "Indian monsoon variability in a global warming scenario", *Natural Hazards*, 29, 189-206.

Krishna Kumar K., Rupa Kumar K., Pant G. B., 1997, "Premonsoon maximum and minimum temperatures over India in relation to the summer monsoon rainfall", *Int. J. Climatology*, 17, 1115-1127.

Liang X., Liu Y. and Wu G., 2005, "The role of land-sea distribution in the formation of the Asian summer monsoon", *Geophysical Research Letters*, 32, doi:10.1029/2004GL021587.

McPhaden M. J., 2004, "Evolution of the 2002/03 El Nino", *Bulletin American Meteorological Society*, 85, 677-695.

Mooley D. A., Parthasarathy B. and Pant G. B., 1986, "Relationship between Indian summer monsoon rainfall and location of the ridge at the 500-mb level along 75°E", *J. Climate Applied Meteorology*, 25, 633-640.

Mooley D. A. and Paolino D. A., 1988, "A predictive monsoon signal in the surface level thermal field over India", *Monthly Weather Review*, 116, 18-37.

Moore J., Grinsted A. and Jevrejeva S., 2006, "Is there evidence for sunspot forcing of climate at multi-year and decadal periods?", *Geophysical Research Letters*, 33, doi:10.1029/2006GL026501.

Pai D. S. and Rajeevan M., 2005, "Empirical prediction of Indian summer monsoon rainfall with different lead periods based on global SST anomalies", *Meteorology Atmospheric Physics*, DOI 10.1007/s00703-005-0136-9.

Rajeevan M., Pai D. S. and Thapliyal V., 1998, "Spatial and temporal relationships between global land surface air temperature anomalies and Indian summer monsoon rainfall", *Meteorology Atmospheric Physics*, 66, 157-171.

Rajeevan M., Pai D. S. and Thapliyal V., 2002, "Predictive relationships between Indian Ocean sea surface temperature and Indian summer monsoon rainfall", *Mausam*, 53, 337-348.

Rasmusson E. M. and Carpenter T. H., 1982, "Variations in tropical sea surface temperatures and surface wind fields associated with the Southern Oscillation/El Nino", *Monthly Weather Review*, 110, 354-384.

Rasmusson E. M. and Carpenter T. H., 1983, "The relationship between eastern equatorial Pacific sea surface temperatures and rainfall over India and Sri Lanka", *Monthly Weather Review*, 111, 517-528.

Reddy P. R. C. and Salvekar P. S., 2003, "Equatorial Indian Ocean sea surface temperature: A new predictor for seasonal and annual rainfall", *Current Science*, 85, 1600-1604.

Robock A., Mu M., Vinnikov K. and Robinson D., 2003, "Land surface conditions over Eurasia and the Indian summer monsoon rainfall", *J. Geophysical Research,* 108, 4131-4147.

Sahai A. K., Grimm A. M., Satyan V. and Pant G. B., 2003, "Long-lead prediction of Indian summer monsoon rainfall from global SST evolution", *Climate Dynamics*, 20, 855-863.

Sankar Rao M., Lau K. M. and Yang S., 1996, "On the relationship between Eurasian snow cover and the Asian summer monsoon", *Int. J. Climatology*, 16, 605-616.

Shukla J., 1987, "Interannual variability of monsoons", *Monsoons* (Ed: Fein J. and Stephens P.), Wiley Interscience, 399-464.

Shukla J. and Paolino D. A., 1983, "The southern oscillation and long-range forecasting of the summer monsoon rainfall over India", *Monthly Weather Review*, 111, 1830 -1837.

Shukla J. and Mooley D. A., 1987, "Empirical prediction of the summer monsoon rainfall over India", *Monthly Weather Review*, 115, 695-703.

Solanki S. K. and Krivova N. A., 2003, "Can solar variability explain global warming since 1970?", *J. Geophysical Research*, A, 108, doi:10.1028/2002JA009753.

Vernekar A. D., Zhou J. and Shukla J., 1995, "The effect of Eurasian snow cover on the Indian monsoon", *J. Climate*, 248-266.

Walker G. T., 1909, "Some applications of statistical methods to seasonal forecasting", [Reprinted in IMS (1986), 43-63.]

Walker G. T., 1923, "Correlation in seasonal variations of weather – VIII. A preliminary study of world weather", *Indian Meteorological Memoirs*, 24, 75-131. [Reprinted in IMS (1986), 120-178.]

Walker G. T., 1924, "Correlation in seasonal variations of weather – IX, A further study of world weather", *Indian Meteorological Memoirs*, 24, 275-332. [Reprinted in IMS (1986), 179-240.]

Wang B., Lin H., Zhang Y. and Lu M. M., 2004, "Definition of South China Sea monsoon onset and commencement of the east Asia summer monsoon", *J. Climate*, 17, 699-710.

Webster P. J. and coauthors, 1998, "The monsoon: processes, predictability and prediction", *J. Geophysical Res.*, 103, 14451-14510.

Wiesnet D. R. and Matson M., 1975, "Monthly winter snowline variation in the northern hemisphere from satellite records 1966-75", *NOAA Tech. Memo., NESS-74*, 19 pp.

Xavier P. K., Marzin C. and Goswami B. N., 2007, "An objective definition of the Indian summer monsoon season and a new perspective on the ENSO-monsoon relationship", *Quarterly J. Royal Meteorological Society*, **133**, 749-76.

# Chapter 3

# Long Range Forecasting of Monsoon Seasonal Rainfall

Predicting the likely physical behaviour of the monsoon system and the resulting distribution of rainfall over different parts of India in different phases of the monsoon season is an extremely challenging task. India is a large country with a multiplicity of climate patterns, and the southwest monsoon season lasts longer than four months. What people ideally want is a calendar of rainfall events to be drawn up well in advance of the arrival of the monsoon so that they can plan and organize their activities during the monsoon season accordingly. This requirement, however, is seemingly impossible to meet.

In order to simplify matters, meteorologists have taken recourse to averaging the monsoon rainfall over the country as a whole and over the entire monsoon season. For ease of computation and comparability of statistics, the period 1 June to 30 September has come to be regarded as the southwest or summer monsoon season, regardless of the fact that the dates of commencement and cessation of monsoon rains differ widely across India. This single average rainfall value, usually abbreviated as AISMR for All-India Summer (or Southwest) Monsoon Rainfall, has been found to serve as a good index of the overall monsoon behaviour for the country and monsoon season as a whole for any year. The process of prediction of the AISMR, leaving out the finer subseasonal or subregional features of the monsoon, is called the 'long range forecasting' of monsoon rainfall.

## 3.1 Interannual Variability of Monsoon Rainfall

The reality, however, is that even if the problem is confined to predicting the AISMR alone, it does not necessarily get simplified, the reason being the nature of the interannual variability of the AISMR. Figure 3.1.1 depicts the manner in which the AISMR has varied between 1875 and 2007, the graph showing the departure from the normal rainfall value of 89 cm in each year during this period.

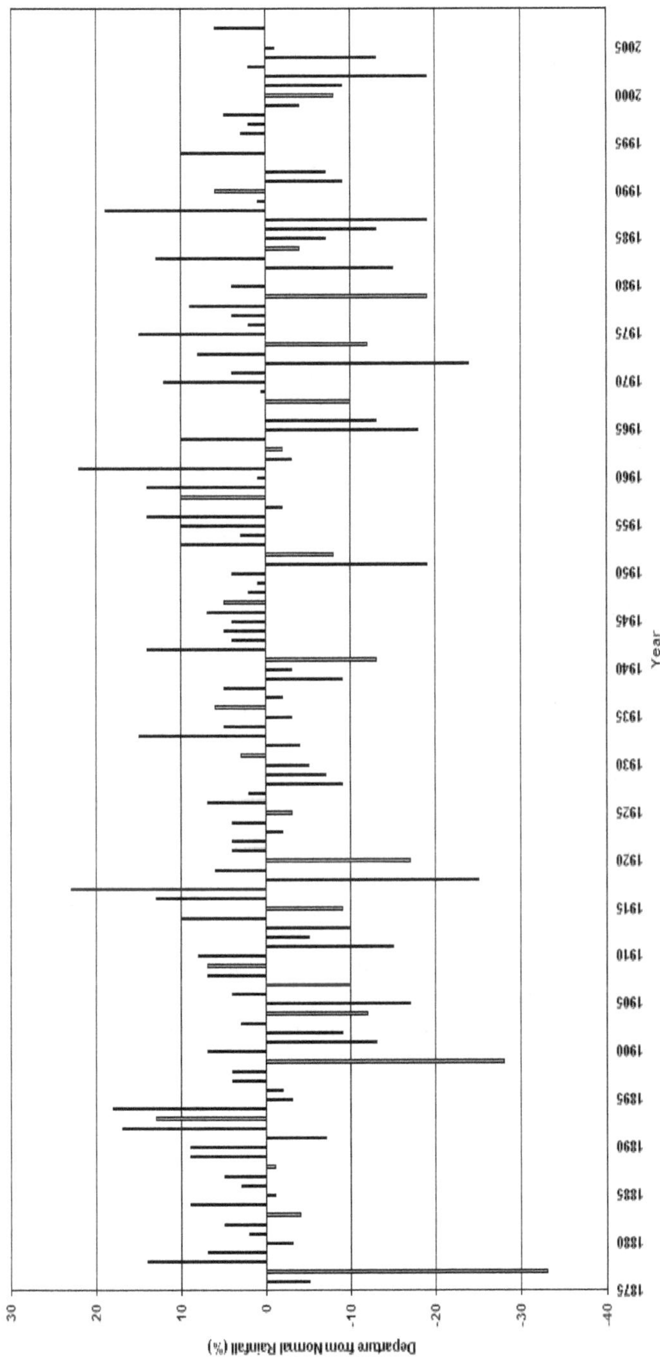


Figure 3.1.1 Interannual variability of All-India Southwest Monsoon Rainfall (Source: IMD)

During the period 1875 to 2007, there were only four years (1877, 1899, 1918 and 1972) when the rainfall deficiency was worse than 20% and just two years (1917 and 1961) when the rainfall excess was over 20%. There have been several years in which the AISMR anomalies were between 10 and 20%, both on the negative and positive sides. However, the graph shows that the years of excess and deficit AISMR have not occurred with any specific or clearly discernable periodicity within these past 133 years. For example, the 43-year period 1921-1964 had only 2 all-India droughts i.e., a higher than 10% deficiency in the AISMR. In comparison, during the 23-year period 1965-1987 there were as many as 8 all-India droughts.

The brighter side of this scenario, however, is that the southwest monsoon is in fact quite regular. It has never happened that it failed to arrive altogether. Its interannual variations have also remained within bounds. Thus the standard deviation of the AISMR is just 10% of the normal.

This dual nature of the interannual variability of the AISMR therefore places two stringent demands on the science and technique of long range forecasting of the AISMR. Firstly, for the long range forecast to be meaningful, it must have a very small error, far less than the standard deviation of 10%. Secondly, the forecasting model must be able to forecast the rainfall extremes, particularly droughts, accurately.

As described in Chapter 2, it has been established since long that the Indian southwest monsoon has strong linkages or teleconnections with several global and regional factors. Such relationships may not be completely understood in a physical sense, but they can be established in terms of statistical correlations. The problem here is that the correlations are not very high, never 100% in any case, leaving room for errors in prediction. Statistical models of long range forecasting are nothing but a possible representation of a composite relationship of the AISMR with several parameters which are individually well-correlated with it. The complexity of such models can therefore vary greatly depending upon how rigorously this composite relationship has been arrived at.

## 3.2 Parametric Models

The simplest of statistical models are those which use the individual correlations of different parameters with the AISMR in a primarily qualitative manner, with just a minimal application of statistics. Such models are called parametric models.

In recent times, the parametric model approach was used successfully by Gowariker et al (1989, 1991) and their model came to be known popularly as the '16-parameter model', after the number of parameters that it employed (Table 3.2.1).

The parametric model of Gowariker et al gave only a qualitative indication of whether the AISMR was likely to be normal (defined as within +10 and -10% of the long period average), deficient (less than -10%) or excess (greater than +10%). The model did not make use of the actual values of the correlation coefficients of the 16 parameters, but only their sign. If a parameter having a positive correlation with the monsoon showed a positive (negative) departure, then it was considered as favourable (unfavourable) to the monsoon. If a parameter having a negative correlation with the monsoon showed a positive (negative) departure, then it was regarded as unfavourable (favourable) to the monsoon.

**Table 3.2.1 Details of the 16 parameters used by Gowariker et al (1991)**

| No. | Parameter | Region | Period |
|---|---|---|---|
| 1 | El Nino | Nino 1+2 | Current year |
| 2 | El Nino | Nino 1+2 | Previous year |
| *3 | Temperature | North India | March |
| 4 | Temperature | Central India | May |
| 5 | Temperature | Northern Hemisphere | November-January |
| 6 | Sea surface temperature | Arabian Sea | January-February |
| *7 | Pressure | Darwin | Spring |
| 8 | Pressure | Northern Hemisphere | January-April |
| 9 | Pressure | Equatorial Indian Ocean | January-May |
| 10 | Pressure | Argentina | April |
| 11 | Southern Oscillation Index | Pacific Ocean | March-May |
| *12 | 10 hPa zonal wind | --- | January |
| *13 | 500 hPa ridge position | India | April |
| 14 | 50 hPa east-west ridge extension | Global | January-February |
| 15 | Snow cover | Himalayan | January-March |
| 16 | Snow cover | Eurasian | December |
| * These parameters were replaced in a revision of the model in 2000 (Table 3.4.1) | | | |

As a thumb rule, established by means of an analysis of past analogues, when more than half of the parameters were favourable in any year, that year's AISMR was in the normal category. Further, whenever more than 70% of the parameters were favourable, the AISMR was on the positive side of the normal. Certain combinations of parameters were also seen to have positive or negative impacts on the monsoon. Starting from the previous winter, as time progressed, the evolution of parameters favourable or unfavourable to the ensuing monsoon could be watched. By analyzing the signals available at any time, a qualitative indication could be obtained from the parametric model about whether the AISMR in the forthcoming monsoon was likely to be in the normal, deficient or excess category. Since several of the 16 parameters required data for the months of April and May, this assessment could best be made at the end of May just before the commencement of the monsoon season.

The application of the parametric model was largely interpretative in nature. Since only the sign of the correlation coefficient had been taken into consideration and not its magnitude, all the parameters had an equal weightage, and the relative strengths of the signals could not be assessed. While the thumb rules were quite clear, they did have some exceptions. Some of the relationships obtained from past data represented conditions that were necessary but not always sufficient. Gowariker et al (1989) claimed that their parametric model was better than previous statistical models, because it had employed a larger number of parameters, and even suggested that their model could be further improved if more parameters could be added. However, they did not address the important issue of how independent parameters could be discovered in a globally connected atmosphere and how far the interdependence of parameters vitiated the performance of a statistical model.

A more recent parametric model is the one constructed by Munot and Krishna Kumar (2007). They have used only six parameters, all of which were derived from NCEP/NCAR reanalysis fields of pressure, temperature and wind at various levels on a $2.5° \times 2.5°$ grid for the period 1949-2005 (Table 3.2.2). The model was run in a hindcast mode. The relationship of each of the six parameters with AISMR was examined critically. Two separate groups comprising 13 years of deficient rainfall and 8 years of excess rainfall were made and the number of parameters which were favourable or unfavourable for the relationship were computed (Table 3.2.3). It was observed that in all the deficient years, four or more parameters contributed to the deficit, while in excess rainfall years, five or all six parameters were favourable for the excess. In all the major drought years like 1972, 1979 and 2002, all of the six parameters were favourable for deficient rainfall.

In comparison, the parametric model of Munot et al (2007) shows a more coherent behaviour than that of Gowariker et al (1991) in which the 16 parameters usually showed a mixed pattern and were never altogether favourable (unfavourable) in a year of excess (deficient) rainfall.

**Table 3.2.2 Details of the 6 parameters used by Munot et al (2007)**

| No. | Parameter | Months (P=Pre-vious) | Region | CC with AISMR |
|---|---|---|---|---|
| p1 | 1000 hPa temperature | PJJA | 20-25N, 62.5-67.5E | +0.51 |
| p2 | 850 hPa temperature | PJJA | 30-35N, 112.5-117.5E | -0.63 |
| p3 | 700 hPa temperature | POCT | 15-20N, 15-25E | +0.65 |
| p4 | 500 hPa zonal wind | MAY | 5-10N, 105-95W | -0.55 |
| p5 | 500 hPa meridional wind | PJJA | 15-25N, 180-170W | -0.60 |
| p6 | 200 hPa meridional wind | PNOV | 10-17.5N, 112.5-117.5E | -0.64 |

**Table 3.2.3 Number of favourable (unfavourable) parameters in excess (deficient) rainfall years in the 6-parameter model of Munot et al (2007)**

| Deficient rainfall | | | Excess rainfall | | |
|---|---|---|---|---|---|
| Year | Number of parameters | | Year | Number of parameters | |
| | Favourable | Unfavourable | | Favourable | Unfavourable |
| 1951 | 5 | 1 | 1956 | 5 | 1 |
| 1965 | 5 | 1 | 1959 | 5 | 1 |
| 1966 | 5 | 1 | 1961 | 6 | 0 |
| 1968 | 6 | 0 | 1970 | 6 | 0 |
| 1972 | 6 | 0 | 1975 | 6 | 0 |
| 1974 | 5 | 1 | 1983 | 6 | 0 |
| 1979 | 6 | 0 | 1988 | 6 | 0 |
| 1982 | 5 | 1 | 1994 | 5 | 1 |
| 1985 | 4 | 2 | | | |
| 1986 | 5 | 1 | | | |
| 1987 | 5 | 1 | | | |
| 2002 | 6 | 0 | | | |
| 2004 | 4 | 2 | | | |

## 3.3 Linear Regression Models

Linear regression analysis has been the most popularly used statistical method for long range forecasting of monsoon rainfall. The problem, however, is that no single predictor has so far been found to have a 100% correlation with the AISMR. In fact, it is difficult to find a predictor that would show a correlation even of the order of 70-80% with the AISMR. This problem has been overcome by the use of several predictors, each having only a moderate value of correlation coefficient of say 50-60%, in a multiple regression scheme. Even in a multiple regression model, the overall correlation coefficient that can be attained is about 95% or so and not 100%, thus always leaving some margin for error in the forecast.

Developing a multiple regression model for forecasting the AISMR is a relatively straightforward matter from the point of view of statistics and models of increasing complexity can be built. The main impediment lies not in the statistical formulation but with the predictors themselves. It is a paradoxical fact that many teleconnections of the southwest monsoon have a sound physical explanation, but their statistical correlation is poor, while many parameters may be found to be highly correlated with the AISMR without an obvious physical cause. The other major difficulty is that the statistical correlations exhibit an epochal behaviour, losing their significance in course of time. The third problem arises from the fact that while the monsoon is a large scale phenomenon involving land, ocean and atmosphere, the predictors are usually found to be highly location-specific as well as time-specific. For example, surface temperature at two adjacent stations may possibly exhibit a strong but opposite correlation with the AISMR. Lastly, while any sound statistical regression model would require the parameters to be independent, in the case of long range forecasting of the monsoon, this condition is difficult to meet, particularly as the number of parameters gets larger. The search for ideal predictors has therefore remained an elusive goal.

From the times of Blanford, Eliot and Walker, until 1960, a variety of predictors including snowfall, sunspot numbers, and river discharges, were tried by IMD in its long range forecasting schemes (Jagannathan 1960). When upper air observations were started over India, these were also used as predictors (Table 3.3.1). As may be seen from the table, most of them had to be eventually given up, barring a few exceptions, while some of them were brought into use again by Gowariker et al (1991).

**Table 3.3.1 Predictors used by IMD for its long range forecast models between 1886 and 1960 (Source: Jagannathan 1960)**

| Predictor | From | To |
|---|---|---|
| *Snowfall:* | | |
| Snowfall | 1886 | 1922 |
| Snowfall conditions June-July | 1897 | 1907 |
| Snow accumulation at end of May | 1906 | 1956 |
| | | |
| *River discharges and floods:* | | |
| Nile floods | 1900 | 1907 |
| Indo-Gangetic River Discharges | 1949 | 1958 |
| Indian river discharges April-May | 1958 | * |
| | | |
| *Sunspot activity:* | | |
| Sunspot periodicity | 1901 | 1904 |
| Phase of sunspot period | 1902 | 1907 |
| | | |
| *Rainfall:* | | |
| Rainfall over east central Africa | 1900 | 1904 |
| Rainfall over south Africa | 1904 | 1907 |
| Rainfall in the equatorial region | 1904 | 1907 |
| Indian rainfall previous year | 1907 | 1919 |
| Subequatorial rain | 1907 | 1919 |
| Zanzibar rain April-May | 1908 | 1936 |
| Ceylon rain May | 1916 | 1924 |
| Java rain (October-February) | 1920 | * |
| Seychelles rain May | 1924 | 1934 |
| Western rain (Baluchistan etc) December | 1924 | 1950 |
| Port Blair rain December | 1924 | * |
| South Rhodesian rain October-April | 1924 | * |
| Quetta rain December | 1950 | * |
| | | |
| *General meteorological conditions:* | | |
| Meteorological conditions over Siberia | 1904 | 1923 |
| Meteorological conditions over west Australia | 1905 | 1907 |
| Meteorological conditions over Chile | 1906 | 1907 |
| Meteorological conditions in northwest India | 1910 | 1924 |
| | | |

| *Pressure:* | | |
|---|---|---|
| Indian pressure previous year | 1907 | 1919 |
| South American pressure March-May | 1907 | 1920 |
| South American pressure April-May | 1920 | * |
| South American pressure December | 1936 | * |
| Mauritius pressure May | 1907 | 1922 |
| Seychelles pressure November-December | 1924 | 1939 |
| Equatorial pressure | 1924 | * |
| Cape pressure September-November | 1924 | 1937 |
| Seychelles pressure November-December | 1924 | 1939 |
| | | |
| *Temperature:* | | |
| Temperature conditions October-November | 1893 | 1910 |
| Dutch Harbour temperature December-April | 1924 | 1956 |
| Punjab temperature range April-May | 1949 | * |
| | | |
| *Surface winds:* | | |
| Strength of southeast trades | 1897 | 1907 |
| Seychelles wind May | 1924 | 1934 |
| | | |
| *Upper winds:* | | |
| Upper data of India | 1920 | 1922 |
| Agra Gwalior 2 km easterly wind March | 1956 | * |
| Calcutta 2 km easterly wind May | 1956 | * |
| Calcutta 4 km easterly wind May | 1956 | * |
| Bangalore 6 km northerly wind April | 1956 | * |
| | * In continuing use in 1960 | |

A typical multiple linear regression model for long range forecasting of monsoon rainfall takes the form of a single equation

$$AISMR = c_0 + c_1 p_1 + c_2 p_2 + \ldots + c_N p_N$$

where $c_0$, $c_1$, $c_2$, .... $c_N$ are the regression constants and $p_1$, $p_2$, .... $p_N$ are the values of the chosen parameters, N being their total number.

A few examples of the regression models are cited below:

A very recent model is that of Munot et al (2007):

$$\text{AISMR} = -7417.7 + 48.87934p_1 - 45.56182p_2 + 23.98553p_3 - 10.45115p_4 - 28.85484p_5 - 21.00083p_6$$

where the 6 parameters are as defined in Table 3.2.2.

Mooley et al (1986) had developed a single parameter model:

$$\text{AISMR} = 38.02 + 3.10x$$

where x is the mean position of the 500 hPa ridge over India along 75° E in the month of April. This relationship was able to explain 53% of the variance in monsoon rainfall.

In an extension of the above work, Shukla and Mooley (1987) developed a 2-parameter model:

$$Y = aX_1 + bX_2$$

where Y, $X_1$ and $X_2$ are the normalized anomalies of AISMR, location of the 500 hPa ridge along 75° E in April, and the change in sea level pressure at Darwin from January to April, respectively.

The regression equation derived by Gilbert Walker on the basis of data from 1865 to 1903, and used by him for the IMD operational long range forecast of the 1909 monsoon rainfall had the following form:

$$\text{Monsoon rainfall} = -0.20p_1 - 0.29p_2 + 0.28p_3 - 1.20p_4$$

where the parameters $p_1$, $p_2$, and $p_4$ were respectively, the snowfall accumulation, Mauritius pressure, and Zanzibar rainfall, all for the month of May, and $p_3$ was the weighted average of pressure for the months March to May, over Argentina and Chile in south America. The AISMR for 1909 was actually in excess of the normal by +7% against +1% predicted by Walker using the above formula (Walker, 1909).


## 3.4 Power Regression Model

In 1906, it was decided by the Government of India to publish IMD's long range forecasts of the monsoon in the Gazette of India and IMD also started preparing a detailed memorandum every year giving the details of the forecast and its verification. Up to the 1930s, IMD's seasonal forecasts used to be worded in somewhat general terms like "rainfall is not likely to be far from normal". After that, the forecasts contained the expected limits within which the rainfall value may lie. From 1949 onwards, two types of long range forecasts were issued every year by IMD, a detailed one for purposes of record and verification, and another one in a simpler language that would be understandable to the public (IMD 1975).

In 1988, IMD adopted two new statistical models for the long range forecast of the monsoon rainfall. One was a parametric model and the other was a power regression model. Both models could be used separately to obtain independent results, but they were related to each other as both models were based upon a common set of parameters. The parametric model initially had a set of 15 parameters (Gowariker et al 1989), but one more parameter was added later (Table 3.2.1) and the model came to be known popularly as the 16-parameter model (Gowariker et al 1991). The parameters were chosen on the basis of their known physical linkages to the monsoon and the strength of their statistical correlations. The 16-parameter model used a tapestry of regional and global parameters that were related to ocean and land temperature, upper winds, surface pressure and snow cover, and belonged to varying periods antecedent to the monsoon. Four of these parameters were later replaced (Thapliyal et al 2003) as their correlations had weakened considerably, but the total number of parameters remained 16 (Table 3.4.1).

**Table 3.4.1 New parameters introduced
in a revision of the 16-parameter model in 2000
(Source: Thapliyal et al 2003)**

| No. | Parameter | Region | Period |
|---|---|---|---|
| 1 | Sea surface temperature | South Indian Ocean | February-March |
| 2 | Temperature | East Coast of India | March |
| 3 | Pressure tendency | Darwin | April to January Difference |
| 4 | Pressure gradient | Europe | January |

India had experienced a drought in 1979 and again in 1982, only to be followed by more monsoon failures in the two consecutive years of 1986 and 1987. In this gloomy agricultural and economic scenario, IMD's forecast of a bountiful monsoon in 1988 was good news that brought cheer to the country, since in a departure from past practice, IMD had gone public with its long range forecast. What added greater credibility to this forecast was the adoption by IMD of the new 16-parameter model. India did in fact receive excellent rains in 1988 as predicted. Thereafter, the country was blessed with a 13-year spell of normal monsoons and year after year, the 16-parameter model correctly predicted a normal monsoon between 1989 and 2001 (Table 3.4.2).

The 16-parameter power regression model (Gowariker et al 1991) was based upon 16 carefully chosen global and regional parameters that had an influence on the monsoon rainfall. Out of these, 6 parameters were related to temperature, 5 to pressure, 3 to wind and 2 to snow cover (Table 3.2.1). The 16-parameter model was used operationally by IMD without any change until 1999. It was only in the year 2000, that a revision was considered unavoidable as some of the statistical correlations were found to have been steadily diminishing over the years (Thapliyal et al 2003). Four of the 16 parameters were replaced by new ones which had better and more stable correlations (Table 3.4.1).

The power regression model was different from linear regression models in that took into account the non-linear interaction of different climate forcings with the Indian monsoon in the following general manner:

$$R = C_0 + \Sigma \ C_i X_i^{Pi}$$

where $R$ is the AISMR, $X_i$ is the value of the $i^{th}$ parameter, $P_i$ is a constant for the $i^{th}$ parameter and has a value between - 4 and +4, $C_0$ and $C_i$ are constants and $\Sigma$ denotes a summation over $i$=1 to 16.

The actual regression equation of the power regression model of Gowariker et al was:

$$\frac{R + \alpha_0}{\beta_0} = C_0 + \sum_{i=1}^{16} C_i \left( \frac{X_i + \alpha_i}{\beta_i} \right)^{p_i}$$

where $\alpha$, $\beta$, $C$ and $p$ are constants so chosen as to produce good forecasts based upon the training data set for the years 1951-1987. Their values were determined by means of a step-by-step iteration and least square fitting. To arrive at the best model several experiments were performed using the parameters in different order, for example in the order of their availability. The final equation was derived on the basis of the parameters being arranged in the descending order of their individual correlation coefficients, which gave the best fit with a model error of 4%. However, Gowariker et al (1989, 1991) did not mention in their papers the correlation coefficients for each individual parameter, nor the correlation coefficient of the combined parameter set, nor the intercorrelations among the parameters themselves.

In a critique of the work of Gowariker et al's work, Delsole and Shukla (2002) commented that the model had 49 independent parameters, as the $\beta$s could be absorbed with the $C$s without loss of generality, leaving a total of three parameters per predictor, plus one constant. Since the 49 parameters were derived from 37 years of data, this model could fit 37 yr of data perfectly and had an artificial skill.

The value of the 16-parameter model did not lie in the discovery of any new regional or global relationships of the monsoon, but in bringing together those already known into a model that could make use of the non-linear part of such relationships. It was therefore claimed to be superior to the many linear multiple regression models which were in vogue at that time, while acknowledging the limitations that apply to any statistical model in general. It was also hoped that the model accuracy could be improved by forming a larger set of parameters.

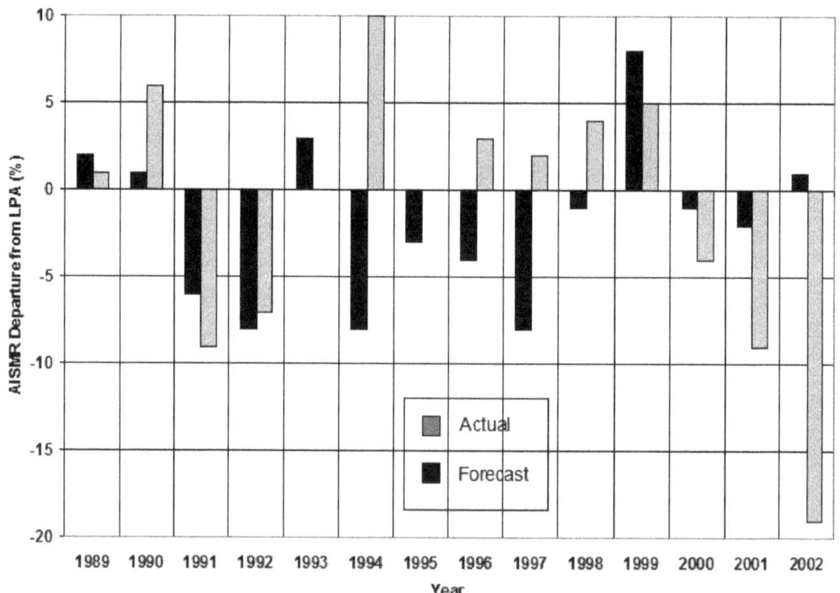


Figure 3.4.1 Comparison of the 16-parameter model forecasts
with the actual AISMR (Source: Gowariker 2002)

**Table 3.4.2 Performance of the 16-parameter model: IMD's operational long range forecasts and actual monsoon seasonal rainfall (Source: Gowariker 2002)**

| Year | AISMR | | | | Deviation (Forecast - Actual) % |
| | IMD long range forecast | | Actual rainfall | | |
| | Category | % of Long Period Average | Category | % of Long Period Average | |
|---|---|---|---|---|---|
| 1988 | Excellent rains* | - | Excellent rains* | - | - |
| 1989 | Normal | 102 | Normal | 101 | +1 |
| 1990 | Normal | 101 | Normal | 106 | -5 |
| 1991 | Normal | 94 | Normal | 91 | +3 |
| 1992 | Normal | 92 | Normal | 93 | -1 |
| 1993 | Normal | 103 | Normal | 100 | +3 |
| 1994 | Normal | 92 | Normal | 110 | -18 |
| 1995 | Normal | 97 | Normal | 100 | -3 |
| 1996 | Normal | 96 | Normal | 103 | -7 |
| 1997 | Normal | 92 | Normal | 102 | -10 |
| 1998 | Normal | 99 | Normal | 105 | -6 |
| 1999 | Normal | 108 | Normal | 96 | +12 |
| 2000 | Normal | 99 | Normal | 92 | +7 |
| 2001 | Normal | 98 | Normal | 91 | +7 |
| *In 1988 no quantitative prediction was made. | | | | | |

The 16-parameter power regression model worked well for the first few years after it was adopted by IMD (Figure 3.4.1 and Table 3.4.2). In 1994, against a prediction of 92%, the actual rainfall was 110% of the long period average, putting the model prediction out by 18%. However, since the actual rainfall was higher than normal, the model behaviour was not taken to be very worrisome. Thereafter, from 1995 to 1998, the model generally underestimated the rainfall. Later this trend reversed, and from 1999 to 2001, the model generally over-estimated the rainfall. The drought of 2002 was totally missed.

## 3.5 Drought Prediction

The process of computation of the AISMR results in masking out the intraseasonal and subregional variations of the monsoon rainfall over India. Therefore the AISMR time series is quite stable and instances of significant departures from the long period average are, thankfully, not that common (Section 3.1). In fact, since 1875, the rainfall deficiency even in poor monsoon years has generally been limited to 20%, there being only a few worse exceptions with higher deficiency: 1877 (-33%), 1899 (-28%), 1918 (-25%) and 1972 (-24%). An all-India drought year is defined as one in which the AISMR is deficient by more than 10% of the normal. By this definition, during the 133 years from 1875 to 2007, there were 20 all-India drought years. These droughts have not been spread uniformly across this time period but they have exhibited a decadal scale variation. Between 1965 and 1987, there were as many as 8 drought years, while from 1921 to 1940, there had been none.

Therefore, on considerations of sheer probability, if one were to just forecast a normal monsoon every year, without using any scientific method or model whatsoever, the chance of such a forecast coming correct would be as high as 80%. Carrying the argument further, if one were to forecast a normal monsoon for a year immediately following an all-India drought year, the probability of its coming correct would be as high as 97%. The basis for this confidence is in the historical data which shows that there were just three past instances of consecutive drought years: 1904-1905, 1965-1966 and 1986-1987. In a training data set of monsoon rainfall that could at best stretch across a hundred and odd years, the sample of drought years becomes relatively quite small, and so any statistical forecast model tends to develop a bias towards forecasting a normal monsoon for the following year. The real test of a statistical forecasting model, or for that matter any model, lies in its ability to foretell the evolution of an upcoming drought.

The 16-parameter model, which had been predicting normal monsoons since its introduction in 1989 until 2001 and proving correct each time at least qualitatively, failed in 2002, when the monsoon failed badly.

### 3.5.1 The Drought of 2002 - Unprecedented and Unpredicted

On 25 May 2002, IMD had issued the long range forecast of southwest monsoon rainfall for the country as a whole and three broad homogenous regions of India, based on the 16-parameter power regression and parametric models. As per this forecast, the rainfall for the southwest monsoon season (June to September 2002) for the country as a whole was likely to be normal, and it was expected that the year 2002 would be the fourteenth normal monsoon year in succession, the normal being defined as rainfall within ±10% of its long period average (LPA). Quantitatively, the monsoon rainfall over the country as a whole was likely to be 101% of its LPA with an estimated model error of ±4%.

The confidence in the forecast of a fourteenth normal monsoon arose from the situation that out of the 16 parameters used by the IMD models, 11 parameters were favourable for a good (normal/excess) monsoon and only 5 were unfavourable (Table 3.5.1.1).

After a normal onset followed by normal rains in the first three weeks of June, conditions soon took an adverse turn. The rainfall during the month of July for the country as a whole, was just half of the normal, the lowest ever recorded. The monsoon revived in August and did not withdraw early, saving the situation from becoming catastrophic. Even then for the June-September season the overall rainfall deficiency was 19%, as many as 61% of the districts received deficient or scanty rainfall, and the country experienced a major drought comparable to the previous droughts of 1987 and 1979.

The overall production of food grains, declined from the record production of 213 million tonnes (mt) achieved in 2001-02 to 175 mt in 2002-03. For the first time, the rabi season production exceeded the kharif production (Figure 1.6.2).

The failure of the 16-parameter model in 2002, after 13 years of continuing success, came in as a rude shock to its designers and users, who had developed a belief in its infallibility, and to the recipients of the forecasts, who had got accustomed to take a normal monsoon for granted. In spite of a spirited defence of the model (Gowariker 2002), there was no escape from the fact that the model had indeed failed.

In an extensive and in-depth survey of the 2002 monsoon, Kalsi et al (2004) have documented various aspects of what they called its 'unusual behaviour', whereas Gadgil et al (2002) described this monsoon as 'intriguing'. This description arose in view of the situation that there was a severe drought for the season as a whole, but the rainfall deficiency for the months of June, August and September was only -4, -4 and -10% respectively. The 2002 drought also did not develop on the lines of past droughts, the monsoon having come in time and not withdrawn early. The three main factors that caused the drought were: (1) the extremely long hiatus in the progress of the monsoon over north and northwest India, resulting in the monsoon finally reaching Rajasthan as late as on 15 August, (2) the unprecedented 51% deficiency of the July rains and (3) the complete absence of monsoon depressions during the entire season. To be fair to the statistical 16-parameter model, it would have been too much to expect it to anticipate these three unusual and intriguing aspects of the 2002 monsoon and account for them in advance. In fact, none of the models which were available at that time, whether operational or experimental, statistical or dynamical, national or international, were able to do so.

In addition to the 16-parameter model that IMD was using to generate its official long range forecasts, IMD had also been running some experimental models in the background. The forecasts from these models were: Dynamic Stochastic Transfer Model 100%, Multiple Regression Model 101%, Power Transfer Model 103% and Neural Network Model 105%.

**Table 3.5.1.1 Parameters favourable and unfavourable for
the 2002 monsoon (Source: Kalsi et al 2004)**

| No. | Parameters | (F) Favourable or (U) Unfavourable |
|---|---|---|
| 1 | El Nino (Same Year) | U |
| 2 | El Nino (Previous Year) | U |
| 3 | South Indian Ocean SST (Feb-Mar) | F |
| 4 | East Coast India Temperature (Mar) | F |
| 5 | Central India Temperature (May) | F |
| 6 | Northern Hemisphere Temperature (Jan-Feb) | F |
| 7 | Arabian Sea SST (Jan-Feb) | F |
| 8 | 50 hpa Wind Pattern (Jan-Feb ) | F |
| 9 | Europe Pressure Gradient (Jan) | F |
| 10 | Southern Oscillation Index (Mar-May) | U |
| 11 | Darwin Pressure Tendency (Apr to Jun) | F |
| 12 | Argentina Pressure (Apr) | F |
| 13 | Northern Hemisphere Pressure (Jan-Apr) | F |
| 14 | Indian Ocean Equatorial Pressure (Jan-May) | U |
| 15 | Himalayan Snow Cover (Jan-Mar) | F |
| 16 | Eurasian Snow Cover (Dec) | U |
|  | Number of Favourable Parameters | 11/16 |
|  | Percentage of Favourable Parameters | 69% |

The neural network model of C-MMACS, Bangalore had given a forecast of 99%. Estimates from several other empirical models were also indicative of an above-average rainfall for the 2002 monsoon season, since most of the parameters had been favourable. The slew of empirical models developed over the years at IITM indicated a consensus forecast of 105% for the season of 2002. The dynamical seasonal forecasts for the 2002 monsoon season issued in May 2002 by the International Research Institute for Climate Prediction (IRI), University of Columbia, had indicated normal monsoon rainfall over India. Only the seasonal predictions of the U. S. National Centre for Environmental Prediction (NCEP) and the European Centre for Medium Range Forecasts (ECMWF) based upon May initial conditions, had indicated possible deficit rainfall over some parts of the country. Only when the June initial conditions were used, did the dynamical models suggest the development of moderate to severe drought conditions (Gadgil et al 2002).

There have been several diagnostic studies into the causes of the drought of 2002. According to Fasullo (2005) the near-equatorial convective anomalies influenced the Hadley and Walker circulation over the Indian Ocean and African region in such a way that there was a reduction of the moisture transport over India. Krishnan et al (2006) have suggested that a dynamical coupling between the Indian Ocean circulation and the southwest monsoon winds may be able to force such drought conditions. Mujumdar et al (2007) have brought attention to the possible linkage of the 2002 drought to the anomalously high convective and cyclonic activity that prevailed over the western Pacific Ocean in that year.

## 3.6 The 8-Parameter and 10-Parameter Models

The year 2002 turned out to be an all-India drought year with an overall rainfall deficiency of 19%, while IMD had predicted a normal monsoon, resulting in a lot of attention being focused on IMD's prediction methodology. Prior to 2002 also, while the overall rainfall was what could be termed as normal, drought situations had been prevailing over many parts of the country, in some areas for successive years. The pressure of user demand had therefore been building up on IMD to issue its forecast much earlier than it did, to give a midseason update, and to provide the forecasts separately for individual months of the monsoon season. There was also scope for narrowing down the IMD's prevailing definition of a 'normal' monsoon which was too broad to bring out patterns of spatial and temporal variability. Besides the failure of the 16-parameter model, these were the factors that motivated IMD, to make an attempt at developing new models with enhanced credibility and which came closer to satisfying the heightened user expectations.

This initiative taken by IMD resulted in the formulation of two new power regression models for monsoon prediction, based upon 8 and 10 parameters respectively. The underlying philosophy and details of methodology have been described in a paper by Rajeevan, Pai, Dikshit and Kelkar (2004) which is hereafter referred as 'RPDK (2004)'. The 8-parameter model was built upon 38 years data (1958-1995) and its independent verification was done for 7 years (1996-2002). It had a model error of 5 % but its main advantage was that it required data only up to March, making it possible to issue a forecast in April itself.

The 10-parameter model was built upon a similar data base and independent verification period. This model used data up to June and had a smaller model error of 4 %, enabling a quantitative midterm update to be generated in July. By that time, the monsoon would have set in and the El Nino trends would be known. Earlier, IMD did not have a model or a system for issuing such an update.

Previously, the users of the IMD long range monsoon forecasts had to wait all along until 25 May as the 16-parameter model required data up to the end of May. Even this necessitated a process of extrapolation of the data available at that time up to 31 May. Furthermore, once IMD issued its forecast on 25 May, there was no system for modifying it as the monsoon advanced into the season.

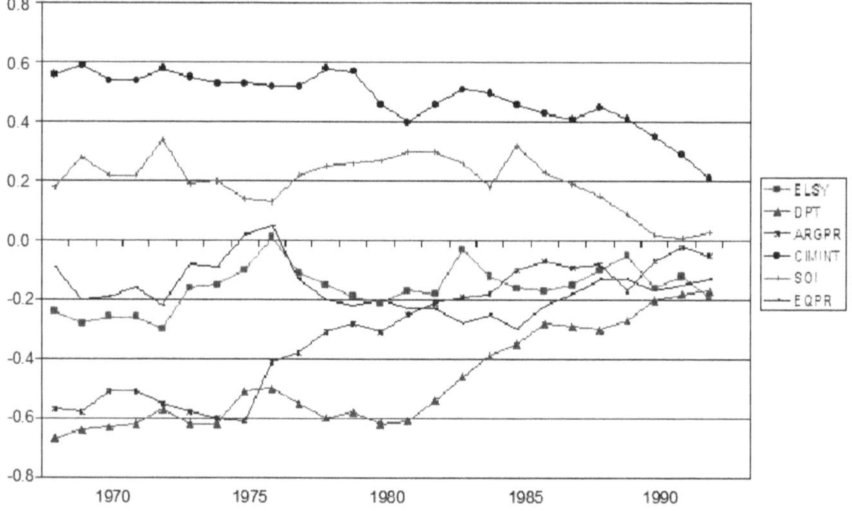



Figure 3.6.1 Moving 21-year correlations with monsoon rainfall of the 6 April-May parameters out of the original 16, showing their weakening with time (Source: RPDK 2004). Legend: ELSY El Nino Same Year, DPT Darwin Pressure Tendency, ARGPR Argentina Pressure, CIMINT Central India Temperature, SOI Southern Oscillation Index, EQPR Equatorial Pressure.

As a part of the development of new power regression models for monsoon rainfall forecasts, every parameter in the 16-parameter model was examined for its statistical stability over time by calculating running 21-year window correlations. Significantly, this analysis revealed progressive weakening of the correlations of all six April-May parameters (Figure 3.6.1) and four winter-spring parameters (Figure 3.6.2). These ten parameters were removed to begin with. An extensive data analysis was also carried out to find more stable and physically related predictors for use in the long-range forecast model. This exercise yielded four new predictors, viz. northwest Europe temperature in January, South Indian Ocean SST gradient index in March, South Indian Ocean zonal wind at 850 hPa in June and Nino 3.4 SST tendency AMJ-JFM. A new parameter set consisting of six old and four new parameters was thus formulated for the purpose of further model development. The 21-year moving correlations of the ten parameters are shown in Figure 3.6.3. All the correlations were found to be stable, especially over recent years.

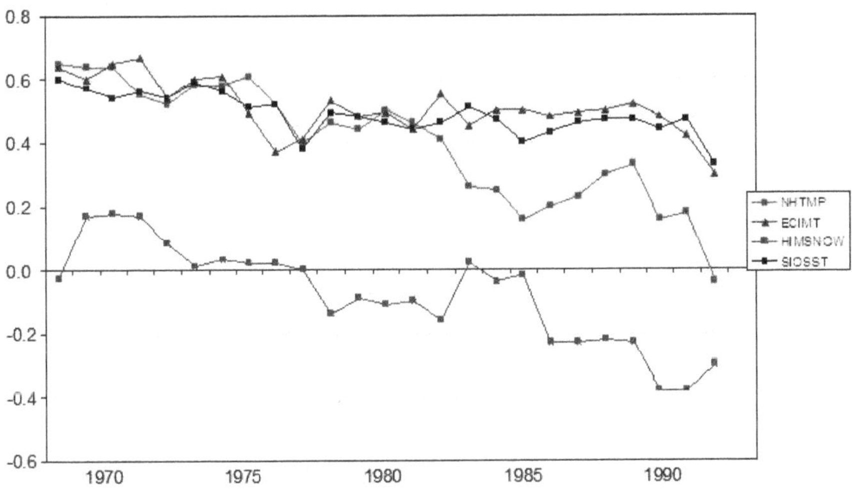


Figure 3.6.2 Moving 21-year correlations with monsoon rainfall of the 4 winter-spring parameters out of the original 16, showing their weakening with time (Source: RPDK 2004). Legend: NHTMP Northern Hemisphere Temperature, ECIMT East Coast India Temperature, HIMSNOW Himalayan Snow Cover, SIOSST South Indian Ocean SST

In the new set of ten parameters, there are eight which become known by March end and two by June end. Using the subset of eight parameters and the full set of ten parameters, the two new power regression models were developed (Tables 3.6.1 and 3.6.2). The predictors were standardized before using them in the power regression equation. The model is nonlinear and the power term $P$ varies between +2 and -2 only. In the 16-parameter model, the power term varied between +4 and -4, which allowed far greater nonlinearity.

The optimum number of predictors in a statistical regression model has always been a matter of debate. There are studies recommending that the predictors be restricted to as small a number as possible in such models. The larger the number of predictors, greater is the chance of their being intercorrelated. So 8 to 10 predictors are required for explaining a good amount of variation (70-75%) in the model development period and also limiting the root mean square error of the results over the independent period to a minimum.

**Table 3.6.1 Details of the new 8-parameter model (Source: RPDK 2004)**

| Model Parameters | Months | C.C. |
|---|---|---|
| El Nino (Previous Year) | Jul-Sep | +0.42 |
| Eurasian Snow Cover | Dec | -0.46 |
| NW Europe Temperature | Jan | +0.45 |
| Europe Pressure Gradient | Jan | +0.42 |
| 50 hPa Wind Pattern | Jan-Feb | -0.50 |
| Arabian Sea SST | Jan-Feb | +0.55 |
| East Asia Pressure | Feb-Mar | +0.61 |
| South Indian Ocean Temperature | Mar | +0.52 |

**Table 3.6.2 Details of the new 10-parameter model**
**(Source: RPDK 2004)**

| Model Parameters | Months | C.C. |
|---|---|---|
| El Nino (Previous Year) | Jul-Sep | +0.42 |
| Eurasian Snow Cover | Dec | -0.46 |
| NW Europe Temp | Jan | +0.45 |
| Europe Pressure Grad | Jan | +0.42 |
| 50 hPa Wind Pattern | Jan-Feb | -0.50 |
| Arabian Sea SST | Jan-Feb | +0.55 |
| East Asia Pressure | Feb-Mar | +0.61 |
| South Indian Ocean SST | Mar | +0.52 |
| Nino 3+4 Temperature | AMJ-JFM Difference | -0.46 |
| South Indian Ocean 850 hPa Zonal Wind | Jun | -0.45 |

The performance of the new 8- and 10-parameter models as compared with IMD's operational forecasts issued with the 16-parameter model during the independent verification period is given in Table 3.6.3. The real test of a statistical model is in its performance during the independent verification period rather than during the model development period or in its hindcast mode. The new 8- and 10-parameter models have generated forecasts that are much closer to actual rainfall than those which were made operationally during the independent verification period. The root mean square error of the operational forecasts during the period 1996-2002 was 11%, while that of the new 8- and 10-parameter models for the same period was 7 and 6% respectively. The model errors of the 8- and 10-parameter models are 5 and 4% respectively, which is of the same order as that of the 16-parameter model at its inception. However, both the new models have not been able to correctly bring out the large rainfall deficiency in 2002, but generally their performance has been better in other drought years in the hindcast mode.

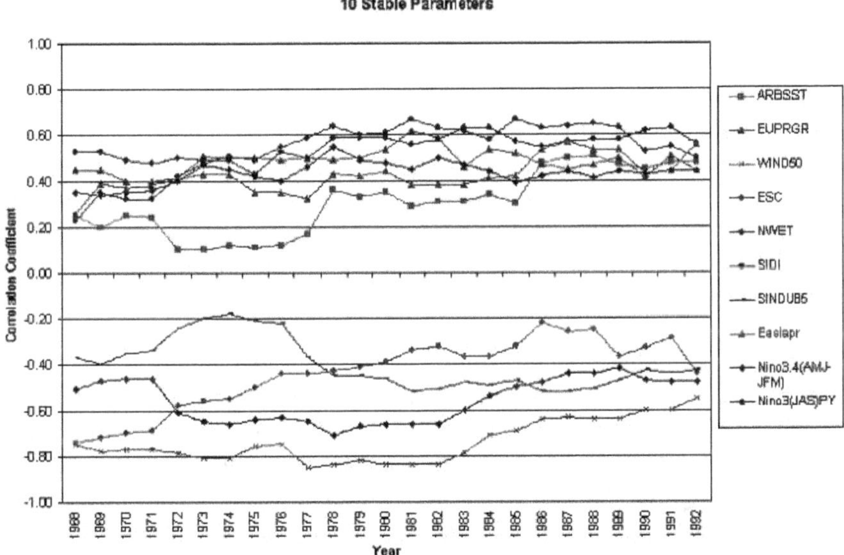


Figure 3.6.3 Moving 21-year correlations with monsoon rainfall of the 10 parameters showing their stability. The year indicates the central year of the 21-year period (Source: RPDK 2004). Legend explained in Table 3.6.2.

**Table 3.6.3 Comparative performance of the new 8- and 10-parameter models and the 16-parameter model (Source: RPDK 2004)**

| Year | Actual - Forecast (%) | | |
|---|---|---|---|
| | 16-Parameter Model | 8-Parameter Model | 10-Parameter Model |
| 1996 | +7 | +3 | +3 |
| 1997 | +10 | -1 | +1 |
| 1998 | +6 | -4 | -2 |
| 1999 | -12 | -5 | -6 |
| 2000 | -7 | +3 | +3 |
| 2001 | -7 | +1 | +3 |
| 2002 | -20 | -17 | -14 |

## 3.7 Forecasting in Terms of Probabilities

While the 16-parameter power regression model (Section 3.4) of Gowariker et al (1991) predicted the departure of monsoon rainfall from the long period average (LPA) in quantitative terms, their parametric model (Section 3.2) provided a qualitative indication of whether the coming monsoon would be normal or not. For purposes of definition, the season's rainfall for the country as a whole between 90 and 110% of the LPA was considered as normal, rainfall below 90% of LPA was regarded as deficient and rainfall above 110% was termed excess.

A look at the interannual variability of the AISMR (Figure 3.1.1) easily reveals that in a climatological sense, the probability of normal/excess rainfall in any given year is about 80%, while the probability of an all-India drought is 20%. This figure of 20% may appear small in comparison but it is far from negligible. The classification of normal, excess and drought into distinct categories also has what could be called 'boundary problems'. It has been the practical experience that the agricultural and economic repercussions of a 90.5% rainfall may be as adverse as those of an 89.5% rainfall situation, but the previous situation would be regarded as normal and the latter as a drought. RPDK (2004) therefore felt that even within the broad category of normal rainfall, there was a need to give a probabilistic indication of the rainfall being in three narrower categories of near normal, below normal and above normal. They accordingly attempted to develop models which can provide the probabilities of rainfall in five different categories ranging from drought to excess rainfall.

These new probabilistic models were based on the linear discriminant analysis (LDA) technique, and the same sets of 8 and 10 parameters (Tables 3.6.1 and 3.6.2) of the power regression models were used. This technique, first introduced by Fischer, is now being commonly used in probabilistic statistical climate prediction (Wilks 1995, Mason et al 2002). Rajeevan et al (2000) have given a detailed explanation of the LDA technique. A prerequisite of the LDA model is for each category to have equal prior probability. In the models of RPDK (2004), the five categories of rainfall mentioned above were so defined as to have equal prior probability of 20% (Table 3.7.1). The models were developed with 40 years of data (1958-97), and data for five years (1998-2002) were used for model verification. First a principal component analysis (PCA) of the 8 and 10 variables was made and then three significant principal components were chosen as predictors. In this case, PC1, PC3 and PC5 were used as predictors to satisfy the relationship between the sample size, number of variables and number of groups.

**Table 3.7.1 Category of monsoon tainfall
for probability-based forecasts (Source: RPDK 2004)**

| Category | Rainfall Range |
|---|---|
| Drought | Less than 90% |
| Below normal | 90-97% |
| Near normal | 98-102% |
| Above normal | 103-110% |
| Excess | Above 110% |

The discriminant model allows the monsoon forecast to be viewed through a different perspective by discriminating between 5 different categories of the rainfall and estimating the probability for each category. In an hindcast mode, the 8-parameter LDA model showed 68% correct classifications, whereas the 10-parameter LDA model showed 78% correct classifications. In hindcast again, both the LDA models correctly gave the highest probability of drought in 8 out of 9 actual drought years during the period, except in 2002 (Table 3.7.2) and no false alarms of drought were generated in any other years. This should be regarded as a significant strength of the probabilistic models.

**Table 3.7.2 Hindcast probability of drought indicated by the LDA model
(Source: RPDK 2004)**

| Year | Probability (%) of Drought Indicated by | |
|------|-----------------------|------------------------|
|      | 8-parameter LDA model | 10-parameter LDA model |
| 1965 | 92 | 96 |
| 1966 | 91 | 78 |
| 1972 | 89 | 99 |
| 1974 | 77 | 66 |
| 1979 | 97 | 99 |
| 1982 | 79 | 89 |
| 1986 | 51 | 55 |
| 1987 | 96 | 99 |
| 2002 | 20 | 4 |

## 3.8 Neural Networks

In recent years, a new statistical modelling technique called the artificial neural networks techniques, has been coming into increasing use as an alternative to the regression technique, whose limitations are otherwise difficult to overcome. In a regression-based prediction model, a functional form like linear, exponential or power, is first chosen and then the coefficients of the equation are derived in such a manner as would minimize the errors. In the neural network (NN) approach, the network is trained with the past data itself to make a functional model. In simple terms, the NN is a coupled input-output map constructed by means of an iterative process. The basic unit of the network is a neuron. As a segment of the input is presented to the network, the skill of the technique is determined by the manner in which the weights of the individual neurons are adjusted in order to predict the next point in the input data series. Once the desired level of accuracy has been reached in the training data set, it can be extended to make a prediction outside the training set.

Goswami et al (1996, 1997) have described the design of a generalized neural network scheme which is more suited to monsoon rainfall prediction. Instead of a simple neuron, they used a composite neuron having non-linear structure functions, which have a greater learning capability.

Guhathakurta et al (1999) argued that the accuracy of a deterministic NN model that is based solely upon the rainfall time series, can be enhanced by including additional signals from the coupled surface-atmosphere system.

For this purpose, they made use of eight regional and global parameters as predictors. Since some of these predictors were intercorrelated, they carried out a principal component analysis of these eight parameters and the first five principal components were taken as inputs to develop a principle component neural network model. Finally they constructed a hybrid NN model which is a simple two-layer NN model without any hidden layer. The input layer of the hybrid model consists of two input nodes which are the outputs of two models, one based on deterministic process that used only the rainfall data series, and the other based on the principal component analysis of eight predictors. Thus the hybrid model has only three unknowns, viz., two weights and a bias, which are obtained by training the NN. Guhathakurta et al (1999) claimed that their hybrid NN model performed better than that of Goswami et al (1997) in predicting the monsoon rainfall. Guhathakurta et al (2006) later applied the NN technique to prediction of monsoon rainfall to the state (subdivision) of Kerala, and its 14 individual districts, and showed that the model performed very well even on these spatial scales (Section 3.12.2).

## 3.9 ARIMA Technique

A statistical technique known as the Auto-Regressive Integrated Moving Average (ARIMA) technique has been found to provide a better alternative to linear regression for the forecasting of rainfall. Delleur and Kavvas (1978) used the ARIMA technique in the modelling of river flow sequences over 14 river basins in the U. S. Their aim at modelling of rainfall sequences was to generate real time statistical forecasts of rainfall, which could serve as inputs to hydrological models. For such purposes it is desirable to have synthetic rainfall sequences which have the same statistical properties as the historical ones but are longer than the duration of the flow sequences. Thapliyal (1981) also demonstrated the usefulness of the leading indicator ARIMA models for seasonal rainfall forecasting over peninsular India.

In a subsequent work, Thapliyal (1982) made an attempt to develop the leading indicator dynamic stochastic model over the peninsular region of India. Such a model utilizes the dynamic relationship between an input $X$ and an output $Y$ of a physical system and disturbance or noise affecting the system. The model thus consists of a dynamic transfer function and a noise transfer function. If the mathematical forms of these two functions are known, an optimal forecast can be made. In the case of a monsoon rainfall model it has to be assumed that the atmosphere is a linear dynamic system which converts the input $X$ into rainfall $Y$. The model output would, however, deviate from the actual rainfall because of the noise or the numerous inputs and feedbacks to the atmospheric system, not all of which are known. By utilizing the dynamics of the ARIMA stochastic model, they can be indirectly incorporated into the model by assuming them as independent shocks or white noise. Thapliyal (1982) used the April mean position of the 500 hPa ridge as the input to develop a dynamic stochastic model for forecasting monsoon rainfall over peninsular India and it was found to perform well.


**3.10 Forecasting by Extension of the Rainfall Time Series**

Had the monsoon been a totally random phenomenon, any attempt to predict it would have been futile. However, it is on the quasi-regularity with which the monsoon appears and behaves every year, that researchers have pinned their hopes of discovering some signal of its future behaviour. There is a school of thought which cherishes the idea that such a signal can be found within the past pattern of monsoon rainfall itself. In other words, the time series of monsoon rainfall is supposed to carry a memory of its own variation and its causes, both known and unknown to us. If this is indeed so, and if a method can be developed to extend this time series, then the future rainfall can be predicted. The difficulty lies in that the rainfall series is far from linear, nor does the curve evolve smoothly, as a major drought may be abruptly followed the very next year by an excellent monsoon, or there could be long spells of normal monsoons. If it is at all possible to design a model which can extrapolate the time series with some mathematical or statistical logic, it needs first to be demonstrated that such a model can successfully replicate the past, so that a rainfall forecast can be made with a sufficient degree of confidence. This is a challenge that has attracted scientists more from outside the meteorological community than within it, for the simple reason that it is difficult for meteorologists to get convinced that past rainfall can somehow predetermine the future rainfall.

In a significant attempt to use this approach for long range forecasting of monsoon rainfall, Iyengar et al (2005) have followed the method of Huang et al (1998) for decomposing time series data into a finite number of empirical modes called Intrinsic Mode Functions (IMFs). Iyengar et al demonstrated that the time series of monsoon seasonal rainfall from 1871 to 1990, for India as a whole, as well as for seven smaller regions, can be decomposed into six statistically uncorrelated modes, the sum of which gives back the original data. What makes this approach more acceptable is that it attempts to provide at least a partial physical insight, by relating monsoon rainfall to meteorological parameters that show similar periodicity as that of a particular IMF.

In the analysis of Iyengar et al, $IMF_1$ is a high-frequency, strongly non-Gaussian mode. It is the most predominant mode that has a period of 2.7 years and makes a contribution of 66% of the interannual variability of all-India monsoon rainfall. $IMF_2$ is the next in importance and it has a period of 5.5 years. The first two modes are probably connected with the quasi-biennial oscillation (QBO) and ENSO phenomena. The IMFs thereafter get progressively slower and less random. $IMF_3$ has a period of 12 years suggestive of an association with the sunspot cycle. The central period of $IMF_4$ is 30 years, which can be related to tidal forcing. $IMF_5$ shows an elongated period of 60 years. $IMF_6$ is the slowest and associated with the climate mode variation over the whole period.

Instead of working with the rainfall data, the IMFs are extrapolated in time through linear regression or neutral network technique, and then added to obtain a forecast of the next year's rainfall. The model was verified on an independent subset of the data series and seen to be efficient enough to capture the drought of 2002, with the help of only antecedent data. To add to their credit, Iyengar et al had forecast a departure of -6% for the monsoon rainfall of 2004, which was better than the official IMD forecast of -2%, the actual departure being -13 %.

There are two significant advantages of extrapolating the rainfall series to get the next year's forecast. The first is that except for the Indian rainfall data which is already well-documented by IMD and IITM, this approach completely eliminates the need for collecting any meteorological or oceanographic data whatever. The process of collection of such diverse data has its own difficulties, but the construction of long term data sets requires several additional aspects to be taken into consideration, such as changes in observational methods over the time period, compatibility of satellite-derived and conventional data, differences between original and reanalyzed data, or even the non-availability of the required historical data. If we deal with rainfall alone, these factors just do not matter and moreover, instead of trying to filter out the noise we are getting rid of the signals themselves!

The second advantage of the approach is in the long lead time that it provides. Since the rainfall time series can be updated immediately after one monsoon season is over, the forecast for the next monsoon can be obtained eight months in advance. The methodology could even perhaps be stretched, with a lesser level of confidence, to forecast the rainfall two monsoons ahead, adding to the advantage (e.g. Goswami et al 1996, 1997).

On the negative side, the main criticism is that the model could not possibly know how the land-atmosphere-ocean system is going to behave between October and the next June or beyond, or what will happen if the system is subjected to a sudden shock that was not anticipated. This is the part that remains unexplained, and hence unacceptable from the strictly meteorological point of view. However, the approach certainly has a potential and needs to be further worked upon.

## 3.11 Ensemble and Projection Pursuit Regression

Rejeevan et al (2006) have described the details of the two new experimental models for the long range forecasting of monsoon rainfall and which were adopted by IMD in 2007 for operational use. These new statistical models are based on the Ensemble Multiple Linear Regression (EMR) and Projection Pursuit Regression (PPR) techniques. While IMD retained its two-stage forecast strategy introduced in 2003, along with the new models, it also changed the set of parameters and further reduced their number.

**Table 3.11.1 Details of predictors used for the first stage forecast (SET-I) by Rajeevan et al (2006)**

| No. | Parameter | Period | Spatial domain | CC with ISMR (1958-2000) |
|---|---|---|---|---|
| A1 | North Atlantic SST anomaly | December -January | 20-30°N, 100-80°W | -0.45** |
| A2 | Equatorial SE Indian Ocean SST anomaly | February -March | 20-10°S, 100-120°E | 0.52** |
| A3 | East Asia surface pressure anomaly | February -March | 35-45°N, 120-130°E | 0.36* |
| A4 | Europe land surface air temperature anomaly | January | Five stations | 0.42** |
| A5 | Northwest Europe surface pressure anomaly tendency | DJF(0) - SON (-1) | 65-75°N, 20-40°E | -0.40** |
| A6 | WWV anomaly | February -March | 5°S-5°N, 120°E-80°W | -0.32* |
| *Significant at and above 5% level | | | **Significant at and above 1% level | |

Two models, EMR-I, and PPR-I, based on a predictor set (SET-I) needing data up to March were developed for the purpose of the first stage forecast issued in April (Table 3.11.1). Two more models, EMR-II and PPR-II, based on another predictor set (SET-II) requiring data up to May were developed for the second stage forecast issued in June (Table 3.11.2). There are a total of 9 predictors as both the sets have three predictors in common. A striking feature of the new models is the elimination of predictors that are close to India, and the inclusion of predictors from the north Atlantic Ocean.

**Table 3.11.2 Details of predictors used for the second stage forecast (SET-II) by Rajeevan et al (2006)**

| No. | Parameter | Period | Spatial domain | CC with ISMR (1958-2000) |
|---|---|---|---|---|
| J1 | North Atlantic SST anomaly | December -January | 20-30°N, 100-80°W | -0.45** |
| J2 | Equatorial SE Indian Ocean SST anomaly | February -March | 20-10°S, 100-120°E | 0.52** |
| J3 | East Asia surface pressure anomaly | February -March | 35-45°N, 120-130°E | 0.36* |
| J4 | Nino-3.4 SST anomaly tendency | MAM(0) –DJF(0) | 5°S-5°N, 170-120°W | -0.46** |
| J5 | North Atlantic surface pressure anomaly | May | 35-°45N, 30-10°W | -0.42** |
| J6 | North Central Pacific zonal wind anomaly at 850 hPa | May | 5-15°N, 180°E-150°W | -0.55** |
| *Significant at and above 5% level      **Significant at and above 1% level | | | | |

This is the first time that the ensemble technique was applied in statistical models for predicting the monsoon rainfall. The use of a weighted ensemble average of a set of suitable selected models with different predictor combinations for the model inferences effectively reduces the error resulting from the use of only a single best model. The PPR is a nonparametric technique, in which the rainfall is expressed as the sum of two parts, linear and non-linear. However, it is not the rainfall series that is decomposed into two parts, but the predictors that are used to predict the linear and nonlinear parts of the rainfall.

The RMS error during the independent forecast period for all the models was relatively smaller than that for the model based on climatology alone. All the models showed better skill and moreover they were also able to correctly forecast the rainfall deficiencies of 2002 and 2004. In general, the performance of the EMR models was better than that of the PPR models.

However, these promising new models did not come up to expectation when used operationally by IMD in 2007. The actual monsoon rainfall was 105% of the long period average, but the IMD long range forecast issued on 19 April 2007 had indicated a rainfall of 95% of the long period average with a model error of ±5%, which was further reduced to 93% with a model error of ±4% in the updated forecast issued on 29 June 2007 (IMD 2007a, 2007b). The reason for the difference between the actual and forecast rainfall is that the model could not anticipate the simultaneous development of Equatorial Indian Ocean anomalies during the monsoon season and the La Nina conditions in September (Bhatia et al 2007).

## 3.12 Forecasting on Finer Spatial and Time Scales

### 3.12.1 Homogeneous Regions

While the prediction of monsoon rainfall over India as a whole has itself been a challenging problem, parallel efforts have always been made to predict the season's rainfall over smaller regions of the country. The motivation arises primarily because India is too large a country, and its geography and topography are such that the pattern of monsoon rainfall is not uniform across it, but differs widely. It has got the wettest places on earth in the northeast, as well as some of the driest places in the northwest. As one goes to smaller spatial domains, the rainfall variability increases on all scales. Therefore, the all-India average rainfall may be a good and useful large scale statistical index of the behaviour of the monsoon in general, but it does not and cannot provide a picture of the rainfall situation prevailing across the country. In fact there have been many years when the all-India rainfall was normal, while hundreds of districts were suffering from rainfall deficiency, as this was compensated in the all-India average by excess rainfall elsewhere. People are not therefore satisfied with just one all-India seasonal rainfall figure and they desire to have forecasts for smaller space and time scales appropriate to practical applications particularly agriculture.

The above considerations and user demands for forecasts of rainfall over smaller regions, has led to the evolution of what are called broad homogeneous rainfall regions of the country. Traditionally, India has been divided into meteorological subdivisions, currently 36 in number, which have a common weather and rainfall pattern within themselves, and are administratively convenient as they are drawn along state and internal boundaries (Figure 1.5.3). Adjacent subdivisions can be clubbed together in many different possible combinations to form a few larger regions. However, a homogeneous rainfall region is usually defined as a group of contiguous subdivisions the rainfall over which is positively and significantly correlated to the area-weighted rainfall of the region covering this group of subdivisions.

In the times of Walker, IMD had been issuing long range forecasts for Burma, northwest India, South Madras (peninsula) and Rest of India or India Main. In 1922, he regrouped these regions as northwest India, northeast India and peninsula but their areas differed from season to season (IMD 1975, IMS 1986).

After a long gap, in 1999, IMD again started issuing long range forecasts for three broad homogeneous regions of India, viz. northwest India, northeast India and the peninsula (Figure 3.12.1.1). The forecasts were generated with three individual power regression models based on three different sets of predictors. In 2003, these were further refined by examining the stability of the parameters, removing the unstable ones and adding some new predictors. The predictors for the three new power-regression models for the three homogeneous regions are given by RPDK (2004) and their model errors were shown to be within ± 8%.

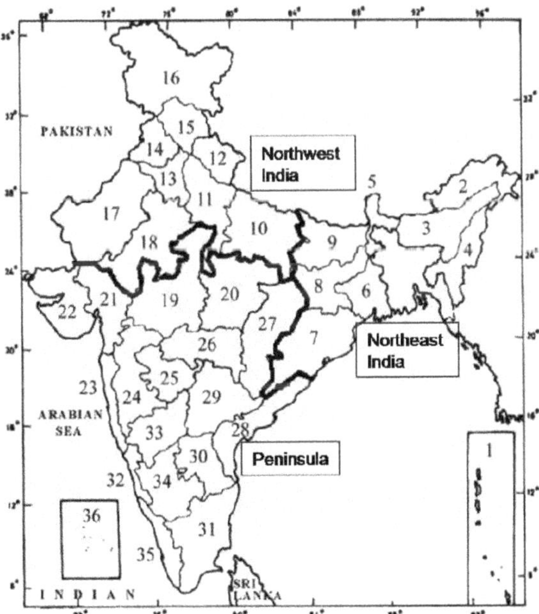


Figure 3.12.1.1 Three broad homogeneous rainfall regions of India: northwest India, northeast India and peninsula (Source: RPDK 2004)

**Table 3.12.1.1 Four homogeneous rainfall regions of India
(Data Source: IMD 2007a, 2007b)**

| Region | Normal monsoon rainfall (mm) | Coefficient of variation (%) | Actual rainfall in 2007 monsoon (mm) | Actual departure (%) | Departure predicted by model (%) |
|---|---|---|---|---|---|
| Northwest India | 611.6 | 19 | 520.8 | -15 | -10 |
| Northeast India | 1427.3 | 8 | 1485.9 | 4 | -2 |
| Central India | 993.9 | 14 | 1073.8 | 8 | -4 |
| South peninsula | 722.6 | 15 | 907.3 | 26 | -6 |

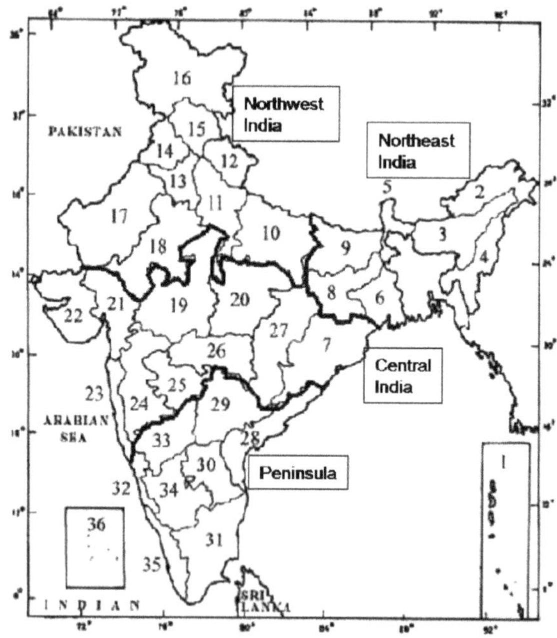



Figure 3.12.1.2 Four broad homogeneous rainfall regions of India: northwest India, northeast India, central India and peninsula (Source: IMD)

**Table 3.12.1.2 Predictors for the multiple regression models for the four homogeneous regions (Source: IMD 2007a)**

| No. | Parameter | Period |
|---|---|---|
|  | Northwest India: |  |
| 1 | North Atlantic msl pressure difference | May |
| 2 | South Atlantic msl pressure | Jan |
| 3 | East Asia msl pressure | Feb-Mar |
| 4 | North central Pacific zonal wind 850 hPa | May |
|  |  |  |
|  | Northeast India: |  |
| 1 | South Atlantic msl pressure | Jan |
| 2 | North Atlantic msl pressure | April |
| 3 | Southeast Pacific SST | Mar |
| 4 | Central Pacific SST | May |
|  |  |  |
|  | Central India: |  |
| 1 | South Indian Ocean SST | Mar |
| 2 | East Asia msl pressure | Mar |
| 3 | North Atlantic SST | Oct-Nov |
| 4 | North Atlantic msl pressure | Mar |
| 5 | Equatorial Indian Ocean msl pressure | Nov |
|  |  |  |
|  | South peninsula: |  |
| 1 | South Indian Ocean SST | May |
| 2 | Southeast Indian Ocean SST | Oct |
| 3 | South Indian Ocean msl pressure | Oct |
| 4 | South Pacific msl pressure | Dec |
| 5 | Arabian 850 hPa zonal wind | Mar |

In 2004, IMD increased the number of homogeneous regions from three to four by splitting peninsular India into central India and south peninsula (Figure 3.12.1.2, Table 3.12.1.1). In 2007, IMD brought new multiple regression models based 4 or 5 predictors, for each of the four regions into operational use (Table 3.12.1.2). These models also have a model error of ± 8%. The performance of these new models for 2007 is given in Table 3.12.1.1. The rainfall over central India and south peninsula was much higher than predicted.

Regression models are, however, not ideally suited for forecasting monsoon rainfall on smaller regions and other methods have been tried with a view to improving the accuracy of the forecasts. Thapliyal (1982) had developed a stochastic dynamic model for predicting the rainfall over peninsular India. In his definition of peninsular India, however, he had omitted Tamil Nadu, Kerala and parts of Karnataka and Andhra Pradesh, comprising the southern peninsula. His model used the location of the 500 hPa ridge over the 75°E longitude (Section 2.6) as the lead indicator and the predictions were close to the observed rainfall.

Rajeevan et al (2000) used three different approaches for forecasting of monsoon rainfall over northwest India and peninsula India, viz., principal component regression, neural network and linear discriminant analysis. All the models showed higher predictive skills over northwest India than over peninsular India, although the three models differed among themselves and had certain advantages and disadvantages.

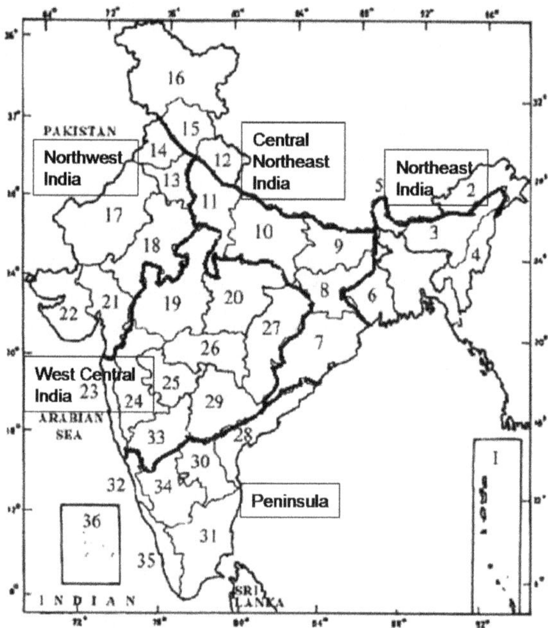


Figure 3.12.1.3 Five broad homogeneous rainfall regions of India: northwest India, central northeast India, northeast India, west central India and peninsula (Source: Kane 2005 and Iyengar et al 2005)

Iyengar et al (2005) used the method of intrinsic mode functions (Section 3.10) to forecast monsoon rainfall over India as well as over smaller regions. For the purpose of their study they divided India into five non-overlapping homogeneous regions: northwest, northeast, west central, central northeast India and peninsula (Figure 3.12.1.3). They also defined another partially overlapping homogeneous region, and two regions were clubbed together to represent what they called the core monsoon region. Their method worked well with the all-India rainfall and also the seven homogeneous regions of the country during independent test periods. Their IMF technique has been discussed in this book earlier (Section 3.10).

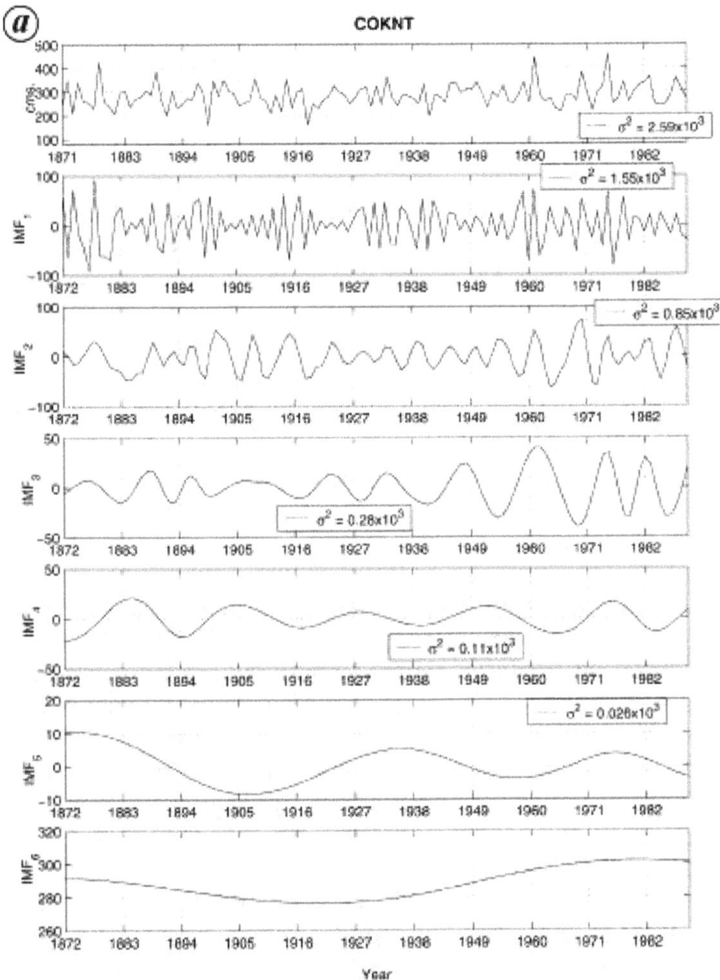

Figure 3.12.2.1 Interannual variation of the six Intrinsic Mode Functions for monsoon rainfall over Coastal Karnataka (Source: Iyengar et al 2006)

Kane (2005) in his study of the correlations of the rainfall over homogeneous regions with ENSO, also used the five regions identical to those of Iyengar et al (2005): northwestern India (6 subdivisions), west central India (9 subdivisions), northeast India (4 subdivisions), central northeast India (5 subdivisions) and peninsula (6 subdivisions). Northwest and west central India together formed what was named as 'homogeneous India' and a combination of west central and part of northwest India was called 'core monsoon India'. The all-India region was made up of all the five homogeneous regions or 30 subdivisions which excluded hilly regions and the islands.

Kane and Iyengar et al have shown that the rainfall is coherent within each of the five different homogeneous regions of India. However, the rainfall data series of the five regions are well-correlated with one another, and a noteworthy point is that some of the correlations are negative. This is why the all-India rainfall series shows a lesser variability in terms of coefficient of variation than the other data series. The rainfall over northwest, west central and central northeast India has a near perfect correlation of 0.99-1.00 with the all-India rainfall. Northeast India has a correlation of 0.92 with the all-India rainfall, while the peninsula has a correlation of only 0.81.

## 3.12.2 Subdivisional Rainfall

The work of Iyengar et al (2005) on forecasting monsoon rainfall for India as a whole and seven smaller regions of the country, using intrinsic mode functions, was further extended by them to the subdivisional scale, with particular reference to North Interior Karnataka, South Interior Karnataka and Coastal Karnataka (Figure 3.12.2.1). Iyengar et al (2006) found that the number and nature of the IMFs computed from the rainfall series for three subdivisions were similar to those for the all-India and regional rainfall. However, on the much smaller spatial scale of a subdivision, the first two IMFs become non-Gaussian because of the increase in rainfall variability. These had to be modelled individually with the help of ANN techniques, while the remaining part was nearly Gaussian and could be modelled as a linear process separately. The method proved successful in eleven out of fourteen years, and it could forecast the persistence of drought-like conditions in South and North Interior Karnataka during 1999-2003. In comparison with regional scale rainfall, the modeling and forecasting efficiency is expectedly lower, and application to other subdivisions is to be investigated.

Guhathakurta (2006) also addressed the issue of the inadequacy of regression-based statistical models for forecasting district-level rainfall and the particular problem of finding large scale predictors when the monsoon rainfall over adjacent districts varied so much, as for example, Kottayam (199.0 cm) and Idukki (269.6 cm) in Kerala. He constructed homogeneous time series for monsoon rainfall for each of the 14 districts of Kerala and the subdivision of Kerala as a whole, using the feed-forward neural network with back-propagation learning algorithm. An attempt was made to predict the monsoon rainfall for these districts separately. Then a prediction for the subdivision of Kerala, was derived as an area weighted average of the districtwise forecasts, which was compared with the forecast made using the subdivisional rainfall time series. Guhathakurta's model showed a good skill in predicting district-level monsoon rainfall over Kerala, which was well within the standard deviation. An interesting result was that the forecast for Kerala obtained by area-weighting the district forecasts was better than that computed for Kerala as a whole.

The two recent studies of Iyengar et al (2006) and Guhathakurta (2006) for Karnataka and Kerala have raised hopes that the spatial limits of regression-based rainfall forecasting can possibly be overcome by means of advanced deterministic techniques like IMFs and neural networks. Such methods have the additional advantage of offering a long time lead of eight months since only the rainfall time series needs to be updated at the end of the monsoon season and no other data are required. Guhathakurta's results suggest that it may be better to first derive forecasts for smaller spatial units and then use them to build the forecasts for a larger area.

### 3.12.3 Monthly Rainfall

Of the four calendar months June to September that are considered for deriving seasonal rainfall statistics of the monsoon, July is the rainiest. In June and September, which are associated with the advance and retreat of the monsoon respectively, the country is only partially covered by monsoon rains. Hence July rainfall accounts for as much as one-third of the season's total for the country as a whole. There is also a good statistical association between July rainfall anomalies and the seasonal (June-September) rainfall anomalies (Gadgil et al 2002). Indian agriculture is most adversely affected by a rainfall deficiency in July, as sowings cannot be delayed any further and crops already sown cannot withstand the acute water stress particularly in the rainfed regions. The severe drought in 2002 was due to the unprecedented rainfall deficiency of -51% in July of that year, which brought down the kharif crop production drastically below that of the previous year (Section 3.5.1).

To develop a model for the agriculturally crucial July rainfall, RPDK (2004) had taken up an extensive analysis of July rainfall variability and associated circulation features. In this process, from a preliminary pool of 15 potential parameters, a set of eight parameters was finally chosen for model development after carefully analyzing their statistical stability (Table 3.12.3.1). An 8-parameter power regression model for July rainfall was developed on the same lines as that for the season's rainfall, using data of 38 years (1958-95) and an independent verification was carried out with data for the subsequent seven years (1996-2002). Although the model error is of the order of ±9%, it is still smaller than the 14% standard deviation of July rainfall, and so the model is useful. The model could not, however, predict in hindsight the large rainfall deficiency of 51% observed in July 2002. This particularly large deficiency of 51% is about three times the standard deviation and can therefore be regarded as a climate extreme, and hence beyond the scope of a statistical model. However, in the case of the other years of deficient July rains, a clear indication of rainfall deficiency could be obtained from the model in the hindcast mode. This model makes use of the Nino 3.4 SST tendency from January to June, as also the South Indian Ocean 850 hPa zonal wind in June, so a forecast of the July rainfall can be issued in early July.

**Table 3.12.3.1 Predictors used in the 8-Parameter multiple tegression model for July rainfall by RPDK (2004)**

| Parameter | Correlation Coefficient (1958-95) |
|---|---|
| South Indian Ocean SST Index Mar-May | 0.39 |
| Nino 3.4 SST Tendency Jan-Jun | -0.30 |
| South Indian Ocean 850 hPa Zonal Wind Jun | -0.45 |
| East Asia Pressure Feb-Mar | 0.30 |
| Eurasian Snow Cover Dec | -0.45 |
| North Indian Ocean-North Pacific Ocean 850 hPa Zonal Wind Difference May | 0.56 |
| Arabian Sea SST Jan-Feb. | 0.40 |
| North Atlantic Ocean SST Dec-Feb | -0.34 |

**Table 3.12.3.2 Predictors Used in the 6-Parameter PCA Model for July Rainfall (Source: IMD 2007a)**

| Parameter | Period |
|---|---|
| North Atlantic SST | Dec |
| Nino 3.4 SST | May-Jun |
| North Pacific msl pressure | Mar |
| East Asia msl pressure | Feb-Mar |
| North Atlantic msl pressure | May |
| Equatorial Indian Ocean msl pressure | Nov |

In 2007, IMD introduced a new model for July rainfall based on principal component analysis technique using 6 predictors (Table 3.12.3.2) which performed well. The model had forecast that the July rainfall was likely to be 95 % of LPA while the actual rainfall was 97 % of LPA (IMD 2007a, 2007b).

## 3.13 Limitations of Statistical Models

In this chapter we have seen both the capabilities and limitations of statistical models for long range forecasting of monsoon rainfall. The regression-based models are basically meant for forecasting the AISMR, but by employing advanced techniques, it is possible to stretch the capabilities of statistical models somewhat to forecast the monsoon rainfall over homogeneous regions or even subdivisions and for certain monsoon months. Even this can be done only at the cost of tolerating higher margins of error and beyond a certain limit the error becomes unacceptable. For forecasting on finer spatial and intraseasonal scales, we have to look towards other means like satellite imagery and numerical models, which will be discussed in the next chapter.

## 3.14 References

Bhatia R. C., Rajeevan M. and Pai D. S., 2007, "An analysis of the operational long range forecasts of 2007 southwest monsoon rainfall", *NCC Research Report No. 5*, India Meteorological Department, Pune, 35 pp.

Delleur J. W. and Kavvas M. L., 1978 "Stochastic models for monthly rainfall forecasting and synthetic generation", *J. Applied Meteorology*, 17, 1528-1536.

Fasullo J., 2005, "Atmospheric hydrology of the anomalous 2002 Indian summer monsoon", *Monthly Weather Review*, 133, 2996-3014.

Gadgil S., Srinivasan J., Nanjundiah Ravi S., Krishna Kumar K., Munot A. A. and Rupa Kumar K., 2002, "On forecasting the Indian summer monsoon: the intriguing season of 2002", *Current Science*, 83, 394-403.

Gadgil S., Rajeevan M. and Nanjundiah Ravi, 2005, "Monsoon prediction - Why yet another failure?", *Current Science*, 88, 1389-1400.

Goswami P. and Srividya, 1996, "A novel neural network design for long range prediction of rainfall pattern", *Current Science*, 70, 447-457.

Goswami P. and Kumar P., 1997, "Experimental annual forecast of all-India mean summer monsoon rainfall for 1997 using a neural network model", *Current Science*, 72, 781-782.

Gowariker V., Thapliyal V., Sarker R. P., Mandal G. S. and Sikka D. R., 1989, "Parametric and power regression models: New approach to long range forecasting of monsoon rainfall in India", *Mausam*, 40, 115-122.

Gowariker V., Thapliyal V., Kulshrestha S. M., Mandal G. S., Sen Roy N. and Sikka D. R., 1991, "A power regression model for long range forecast of southwest monsoon rainfall over India", *Mausam*, 42, 125-130.

Gowariker V., 2002, "Reflecting on IMD's forecast model", *Current Science*, 83, 936-938.

Guhathakurta P., Rajeevan M. and Thapliyal V., 1999, "Long range forecasting Indian summer monsoon rainfall by a hybrid principal component neural network model", *Meteorology Atmospheric Physics*. 71, 255-266.

Guhathakurta P., 2006, "Long-range monsoon rainfall prediction of 2005 for the districts and sub-division Kerala with artificial neural network", *Current Science*, 90, 773-779.

Huang N. E. and coauthors, 1998, "The empirical mode decomposition and the Hilbert spectrum for nonlinear and nonstationary time series analysis", *Proc. Royal Society*, London, A454, 903-995.

IMD, 1975, *Hundred Years of Weather Service (1875-1975)*, India Meteorological Department, Pune, 207 pp.

IMD, 2007a, *Memorandum on update of long range forecast for southwest monsoon*, India Meteorological Department, New Delhi, 29 June 2007.

IMD, 2007b, *Southwest monsoon 2007 end-of-season report*, India Meteorological Department, New Delhi, 5 October 2007.

IMS, 1986, *Sir Gilbert Walker Selected Papers, Long Range Forecasting of Monsoon Rainfall*, Indian Meteorological Society, New Delhi, 254 pp.

Iyengar R. N. and Raghu Kanth S. T. G.. 2005, "Intrinsic mode functions and a strategy for forecasting Indian monsoon rainfall", *Meteorology Atmospheric Physics*, 90, 17-36.

Iyengar R. N. and Raghu Kanth S. T. G., 2006, "Forecasting of seasonal monsoon rainfall at subdivision level", *Current Science*, 91, 350-356.

Kalsi S. R. and coauthors, 2004, "Various aspects of unusual behaviour of monsoon 2002", *Meteorological Monograph No. Synoptic Meteorology 2/2004*, India Meteorological Department , Pune, 105 pp.

Kane R. P., 2005, "Unstable ENSO relationship with Indian regional rainfall", *Int. J. Climatology*, 25, DOI: 10.1002/joc.1281.

Krishnan R. and coauthors, 2006, "Indian Ocean-monsoon coupled interactions and impending monsoon droughts", *Geophysical Research Letters*, 33, doi:10.1029/2006GL025811.

Mason S. J. and Mimmack G. M., 2002, "Comparison of some statistical methods of probabilistic forecasting of ENSO". *J. Climate*, 15, 8-29.

Mooley D. A., Parthasarathy B. and Pant G. B., 1986, "Relationship between Indian summer monsoon rainfall and location of the ridge at the 500-mb level along 75° E", *J. Climate Applied Meteorology,* 25, 633-640.

Mujumdar M., Vinay Kumar and Krishnan R., 2007, "The Indian summer monsoon drought of 2002 and its linkage with tropical convective activity over northwest Pacific", *Climate Dynamics*, 28, 743-758.

Munot A. A. and Krishna Kumar K., 2007, "Long range prediction of Indian summer monsoon rainfall", *J. Earth System Science*, 116, 73-79.

Rajeevan M., 2001, Prediction of Indian summer monsoon: Status, problems and prospects", *Current Science,* 81, 1451-1457.

Rajeevan M., Guhathakurta P. and Thapliyal V., 2000, "New models for long range forecasts of summer monsoon rainfall over north west and peninsular India", *Meteorology Atmospheric Physics*, 73, 211-225.

Rajeevan M., Pai D. S., Dikshit S. K. and Kelkar R. R., 2004, "IMD's new operational models for long range forecast of southwest monsoon rainfall over India and their verification for 2003", *Current Science*, 86, 422-431. [Referred to in this chapter by the abbreviation RPDK (2004).]

Rajeevan M., Pai D. S., Anil Kumar R. and Lal B., 2006, "New statistical models for long-range forecasting of southwest monsoon rainfall over India", *Climate Dynamics*, DOI 10.1007/s00382-006-0197-6.

Sikka D. R., 2003, "Evaluation of monitoring and forecasting of summer monsoon rainfall over India and a review of monsoon drought of 2002, *Proc. Indian National Science Academy*, 69, 479-504.

Shukla J. and Mooley D. A., 1987, "Empirical prediction of the summer monsoon rainfall over India", *Monthly Weather Review,* 115, 695-703.

Thapliyal V., 1981, "ARIMA model for long range prediction of monsoon rainfall in peninsular India", *Meteorological Monograph No. Climatology/12,* India Meteorological Department, Pune, 12 pp.

Thapliyal V., 1982, "Stochastic dynamical model for long range prediction of monsoon rainfall in Peninsular India", *Mausam*, 33, 399-404.

Thapliyal V. and Rajeevan M., 2003, "Updated operational models for long-range forecasts of Indian summer monsoon rainfall", *Mausam*, 54, 495-504.

Walker G. T., 1909, "Some applications of statistical methods to seasonal forecasting" [Reprinted in IMS (1986), 43-63.]

Webster P. J. and Yang S., 1992, "Monsoon and ENSO: Selectively interactive systems", *Quarterly J. Royal Meteorological Society*, 118 (B), 877-925.

Wilks D. S., 1995, *"Statistical methods in the atmospheric sciences"*, Academic Press, 467 pp.

# Chapter 4

# Prediction of Monsoon Intraseasonal Elements

The various statistical techniques for long range forecasting of the all-India monsoon seasonal rainfall have been dealt with elaborately in the previous chapter. Attempts to stretch the techniques further to forecast rainfall over a few smaller homogeneous regions of the country, or even for certain subdivisions, and on smaller time scales like a month, have also been discussed.

While the AISMR is a gross index of the monsoon rainfall on large scale considerations, it is easily understandable that different combinations of positive and negative rainfall anomalies on the smaller scale can possibly lead to the same average value of the AISMR. Large parts of the country may be reeling under drought conditions while there may be floods in other parts due to excess rains, but this may get evened out in the computation of the AISMR and produce an average value that is close to normal. In the opposite case, the AISMR may be indicative of a deficient monsoon, but the rain spells might have been received optimally just when the crops required them, and the adverse impact on agriculture would not then be very perceptible.

The prediction of AISMR cannot therefore serve as a proxy for the prediction of monsoon rainfall on the subregional and intraseasonal scales. The country's agricultural and water management strategy evolves in real time in response to the following elements of the monsoon and it is important that they should be properly predicted.

*Date of onset of the monsoon over Kerala:* Unless the monsoon enters Kerala, it cannot progress ahead, at least over the peninsula. The mean date of onset over Kerala is 1 June with a standard deviation of a week. A delayed onset means a prolonged summer, and postponed sowings. Farmers suffer huge losses if they do the sowings, but the rains do not sustain thereafter.

*Hiatus in the progress of the monsoon:* If after the onset over Kerala, the monsoon does not advance further at the normal speed, farmers in the interior parts of the country have similar problems with their sowing operations.

*The axis of the monsoon trough:* For a stable monsoon circulation pattern, the monsoon trough must get established in its normal position.

*Monsoon depressions:* These systems form over the Bay of Bengal and then move inland. The paths which they follow determine which parts of the country will get copious rains and which rivers will get flooded. The number of lows and depressions in the season has an important bearing on the AISMR.

*Dry and wet spells:* Crops need alternate dry and wet spells for optimum growth. Prolonged dry spells lead to water stress in crops and may even lead to local drought conditions.

*Breaks in the monsoon:* When the monsoon trough moves to the Himalayan foothills, monsoon activity is at its lowest and this is called a break situation. Persistent and intense breaks may result in all-India drought.

*Flash floods:* Heavy rainfall in river catchment areas cause downstream flooding, particularly in hilly regions.

*Date of withdrawal of the southwest monsoon:* If the monsoon withdraws earlier than normal, crops may suffer water stress and the yields may get reduced, while reservoirs may not get filled to capacity. A delayed withdrawal is beneficial from this viewpoint.


## 4.1 Onset of the Southwest Monsoon over Kerala

While it is relatively easy to define the onset of the monsoon in a meteorologically precise manner, it is a very difficult task to arrive at a definition that is also meaningful to the general public. The onset of the southwest monsoon over Kerala has been the subject of numerous investigations (Joshi et al 1994, Joseph et al 1994, 2003, Khole et al 2004, Pearce et al 1984, Shenoi et al 1999, Soman et al 1993). There are several characteristic meteorological features which are associated with the onset over Kerala, and rainfall is only one of them. However, it is not necessary that all of these features must be present at the time of the onset every year, and the monsoon can arrive even without some of the conditions being fulfilled. Further, there are some meteorological scenarios which resemble an onset very closely, but they do not produce the sustained increase in rainfall that people expect from the monsoon. This is a particularly important consideration for farmers who have to start their sowing operations for kharif crops. The evolution of a commonly acceptable definition of the onset has therefore remained an elusive and challenging goal.

Ramesh et al (1996) and Basu et al (1999) have defined objective norms for the definition of the onset, based upon the NCMRWF forecast fields of wind, temperature and humidity at ten pressure levels over the monsoon region. The intensification of the monsoon flow is very sensitive to the tropospheric diabatic heating over the Arabian Sea. Hence for determining the onset, they selected three factors for monitoring from the first week of May: net tropospheric moisture buildup, mean tropospheric temperature increase, and sharp rise in the 850 hPa kinetic energy over the Arabian Sea.

Figure 4.1.1.1  Kalpana-1 infra-red channel image of 26 May 2006 showing
an early and strong onset of the southwest monsoon over Kerala
(Source: IMD)

The classical definition of the monsoon onset over Kerala was based primarily on rainfall (Ananthakrishnan et al 1967, 1968) as satellite imagery and numerical model predictions were not available at that time. A detectable increase in the pentad rainfall over selected stations in Kerala was seen to be a good indicator of the onset of the monsoon. While this approach worked well for most of the onsets, it failed to give an objective guidance in those years in which strong premonsoon thunderstorm activity continued right up to the date of the monsoon onset. IMD's operational declarations of the onset of monsoon over Kerala in the past have had to be based upon all such considerations and some amount of subjectivity could not have been avoided.

In the most recent work on this subject, Joseph et al (2006) have used NCAR/NCEP reanalysis data on wind and integrated water vapour for the years 1971-2003, supplemented by NOAA OLR data and SST derived from TRMM TMI, with which they have formulated a scheme for defining the onset date objectively. Their scheme does not require rainfall data over Kerala explicitly, OLR being used as a proxy measure of convection. Interestingly, the onset dates reworked objectively in this manner, do not differ significantly from IMD's operationally declared onset dates except in a few years.

Figure 4.1.1.2  Kalpana-1 visible channel image of 5 June 2005 showing a weak and delayed onset of the southwest monsoon over Kerala
(Source: IMD)

In 2006, IMD adopted a new set of criteria (Pai et al 2007) which can be applied objectively for the declaration of the monsoon onset over Kerala (Table 4.1.1).

### Table 4.1.1 New IMD criteria for declaring the monsoon onset over Kerala (Source: Pai et al 2007)

| | Parameters | Criteria |
|---|---|---|
| 1 | Rainfall recorded at 14 stations in and around Kerala: Minicoy, Amini, Thiruvananthapuram, Punalur, Kollam, Allapuzha, Kottayam, Kochi, Trissur, Kozhikode, Talassery, Cannur, Kasargode and Mangalore | 60% of available stations should report rainfall of 2.5 mm or more on two consecutive days, after 10 May |
| 2 | Depth of lower tropospheric westerlies over the region 0-10° N and 55-80° E | Depth of westerlies should be maintained up to 600 hPa level |
| 3 | Strength of zonal wind over the region 5-10° N and 70-80° E | Wind speeds should be 15-20 knots at 925 hPa level |
| 4 | INSAT-derived OLR over the region 5-10° N and 70-75° E | OLR should be less than 200 w m$^{-2}$ |

In these new criteria, there is still an emphasis on the detection of the sharp increase of rainfall over Kerala that characterizes the onset. However, they now take into consideration whether the associated large scale circulation features have been established or not, on the lines of Joseph et al (2006), the data required being obtained from the RSMC wind analysis or satellite-derived cloud motion vectors. This is a good attempt towards unifying the various definitions based only on rainfall or on the circulation features alone.

With the onset dates worked out on the basis of the revised criteria for the years 1971-2007, the mean onset date remains at 1 June, and the standard deviation at 8 days. During this period, the earliest onset was on 18 May in the year 1990 and the most delayed onset was on 19 June in 1972.

### 4.1.1 Forecasting the Date of Onset over Kerala

The best precursors of monsoon onset over Kerala are observed in satellite imagery such as that provided by INSAT. These images have clear indications of whether the onset is early and strong (Figure 4.1.1.1) or whether it is delayed and weak (Figure 4.1.1.2). In many years, INSAT images show convective clouds along the west coast of India, indicative of a trough in the westerly winds over the Arabian Sea. Water vapour channel imagery vividly brings out the onset over Kerala and shows the distinctly moist air of the monsoon with dry regions to its south and north (Figure 4.1.1.3). Weekly data of Oceansat-1 MSMR for the premonsoon season of the year 2000 showed that as the monsoon progressed towards the Indian subcontinent, the area of maximum moisture content and wind speed moved northwards. There was a sharp increase in water vapour over western Indian Ocean about three weeks prior to the onset over Kerala and a rise in the wind speed over western Arabian Sea just prior to the onset (Simon et al 2001).

Figure 4.1.1.3 Kalpana-1 full disc water vapour channel image of 27 May 2006 showing the onset of the southwest monsoon and the distinctly moist air of the monsoon with dry regions to its south and north (Source: IMD)

According to Joseph et al (2006), on an average, eight pentads before the onset of the monsoon over Kerala, a spatially large area of deep convection builds up near the equator, south of the Bay of Bengal. This is the first advance signal of the monsoon onset process and it is possible to see this development in satellite imagery (Figure 4.1.1.4).

This is followed by the formation of another area of convection also near the equator but south of the Arabian Sea, on an average, three pentads before the onset over Kerala. This heat source and the associated crossequatorial low level jet stream (LLJ) strengthens while moving northward and passes through Kerala at the time of the monsoon. Although Joseph et al have used these criteria to objectively define the date of onset of monsoon over Kerala, they could as well be used to predict the date of onset eight to three pentads in advance, with a proper safeguard against the detection of onsets that could later turn out to be spurious.

Figure 4.1.1.4 Kalpana-1 infra-red channel image of 21 April 2008 showing a buildup of convection as an early precursor to the monsoon onset over Kerala (Source:http://www.imd.gov.in)

Pai et al (2007) have developed two principal component regression models for forecasting the onset date over Kerala. This type of models are better than simple multiple regression models when there is some intercorrelation amongst the predictors, as is the case here. The need for two models arises because while the mean onset date is 1 June, there is some possibility of the onset occurring much earlier in May. The two models make use of data for the second half of April and first half of May respectively, with a total of 9 parameters (Table 4.1.1.1). The premonsoon rainfall peak which is used as a predictor, is an event that occurs 6-8 pentads ahead of the onset and is referred to in the above paragraph. It is associated with the passage of a cloud band across Kerala similar to that at the onset but without deep convection prevailing over southeast Arabian Sea.

The models during independent verification, have yielded forecasts of the onset date which are on an average within 3 to 4 days of the actual onset date, but the two models themselves differ by 2 to 3 days. However, the performance is better than that of purely climatology-based models.

**Table 4.1.1.1 Predictors for onset of monsoon over Kerala**
**(Source: Pai et al 2007)**

| | Predictor | Period | Region | C. C. (%) (1975-2000) |
|---|---|---|---|---|
| 1 | Minimum surface air temperature | | Northwest India | -0.38 |
| 2 | MSL pressure | | Subtropical northwest Pacific | 0.57 |
| 3 | Zonal wind at 925 hPa | 16-30 April | Northeast Indian Ocean | -0.52 |
| 4 | Zonal wind at 200 hPa | | Indonesian region | 0.48 |
| 5 | OLR | | South China Sea | 0.40 |
| 6 | Premonsoon rainfall peak | April-May | South peninsular India | 0.48 |
| 7 | Minimum surface air temperature | | Northwest India | -0.37 |
| 8 | Zonal wind at 925 hPa | 1-15 May | Equatorial south Indian Ocean | 0.52 |
| 9 | OLR | | Southwest Pacific | -0.53 |

## 4.1.2 Other Onset-related Considerations

*Onset vortex*: The formation of the so-called onset vortex over southeast Arabian Sea is a significant element of the onset process. Krishnamurti et al (1981) made one of the first detailed examinations of the onset vortex with MONEX-79 and other data. One of their findings was that the kinetic energy of the zonal flow over the central Arabian Sea increases by an order of magnitude one week prior to the commencement of monsoon rain over central India.

Soman et al (1993) constructed composite fields of various atmospheric variables relative to a uniform set of onset dates. The onset vortex that takes the monsoon northward along the west coast in many years was clearly discernible between 600 and 400 hPa in their composite streamline charts.

Ocean observations reveal that during the premonsoon months of February-May, the near-surface waters in the Arabian Sea progressively warm up to form a mini-warm pool with a core of SSTs higher than 30 °C in the southeastern region. While examining the physical mechanism behind this development, Rao et al (1999) observed that this mini-warm pool is the most favoured region for the genesis of the onset vortices on most occasions. Their conclusion is based upon the locations of SST maxima in individual years between 1961 and 1990.

The formation of the onset vortex has been analysed through a numerical model by Deepa et al (2007). They examined the SST distribution over the Arabian Sea and circulation at 850 hPa to identify the positions of the mini-warm pool and the low level jet around the onset date for six years 2000-2005. The study revealed that the onset vortex had formed only in 2001 under the influence of the mini-warm pool on the northern flank of the low level jet. During other years it seldom formed due to the absence of a mini-warm pool, a weak low level jet and a lack of shear. A review of the current understanding of the Arabian Sea mini-warm pool and its role in the formation of the monsoon onset vortex has been presented by Vinayachandran et al (2007).

*Advance of the monsoon over northeast India*: There are no well-defined criteria for defining the onset of the monsoon over the Andaman Sea. Climatologically speaking, the onset over the Andaman Sea takes place prior to the onset over Kerala. In some years, the Bay of Bengal branch of the southwest monsoon makes a more rapid northward advance. This results in the monsoon entering northeast India even before it has made its official entry into Kerala. Many times, the monsoon makes its appearance over northeast India almost in parallel. Here again, premonsoon thunderstorm activity is common and it could be difficult to make a clear distinction between premonsoon and monsoon rains based upon rainfall figures alone.

*Tropical cyclones in the onset phase*: The development of tropical cyclones in the last week of May or early June and the path traversed by them, significantly modifies the onset pattern of the monsoon. A storm forming at this time over the Bay of Bengal may take the monsoon along with it towards northeast India ahead of the normal date. On the other hand, a storm that has formed over the Arabian Sea may move away from the Indian mainland and disturb the onset process which has then to reorganize itself after the storm has dissipated.

In 2007, a cyclonic storm 'Akash' formed over east central Bay of Bengal on 13 May, then moved northeastward and crossed the Bangladesh coastline. It affected the normal flow and there was an anomalous intrusion of midlatitude westerlies into the region. The monsoon onset over Kerala took place on 28 May, four days ahead of the normal date, but only to be followed by the formation on 1 June by another storm, supercyclone 'Gonu' over east central Arabian Sea. It moved northwestward and crossed the coast of Oman. Both these systems, which formed in the onset phase, disrupted the normal monsoon flow. After the onset over Kerala, the monsoon started to move further only after 9 days. In the end, however, the monsoon reached west Rajasthan on 4 July, 11 days before its normal date of 15 July.

### 4.1.3 Advance of Monsoon into Interior Parts of India

Once the monsoon onset has occurred over Kerala, it does not seem to have any relevance to the monsoon's advance into the interior parts of India (Figure 1.4.1) which is governed by other factors. The monsoon covers the whole country by 15 July on an average, but the speed of advance has considerable variation along the way.

Once the onset over Kerala has been declared operationally, it is a meteorological practice to draw a line on the surface weather map that represents the northern limit of the monsoon (NLM). South of the NLM lies the area that has been covered by monsoon rains and the NLM is moved progressively northward as the monsoon advances into the country. Here again there are no objective criteria for determining the NLM, and it is mostly done qualitatively with rainfall as a prime consideration. Ramesh et al (1996) and Basu et al (1999) have used two criteria for determining the NLM. One is the steady increase of total precipitable water over a station under the climatologically expected wind regime, and the other is the fall and subsequent rise in the net moisture content under the same wind regime. If these conditions are satisfied under a different wind regime such as a westerly trough, then the rainfall is considered as premonsoon.

## 4.1.4 Hiatus in the Advance

After a timely or early onset over Kerala, the monsoon may linger on somewhere on the way in what is called a hiatus and it may reach many regions later than normal. Or, in spite of a late onset over Kerala, the monsoon advance can accelerate and reach northwest India ahead of its normal date.

The monsoon of 2002 had three hiatus epochs during its advance into north India. The first one was from 13 to 19 June over central India, the second from 5 to 18 July over west Uttar Pradesh and east Rajasthan, and the third from 20 July to 14 August over west Rajasthan. The third hiatus was almost in continuity and an extension of the second (Kalsi et al 2004). This situation resulted in a major failure of the July rains and a drought for the season as a whole (Section 3.5.1)

## 4.2 Intraseasonal Variability of Monsoon Rainfall

In one of the earliest studies of the monsoon based upon satellite imagery, Sikka and Gadgil (1980) studied daily variations of the maximum cloud zone (MCZ) and the 700 hPa trough over the region covering the Indian longitudes 70-90° E from the equator to 35° N in the months of April to October during the years 1973-1977. They observed that in the monsoon months of June-September, two MCZs exist in the region, one over the continent and another over the ocean. Both the MCZs exhibit alternating episodes of strong and weak intensity, but the strength of the two MCZs is oppositely correlated. Both the MCZs show a northward propagation, and there is a period of 3-4 weeks between successive generations of the northward moving cloud bands.

## 4.2.1 Oscillations and Modes

The Madden-Julian Oscillation (MJO) is an intraseasonal oscillation that occurs in the global tropics and affects many atmospheric and oceanic parameters like wind, cloudiness, rainfall, SST and ocean evaporation. The MJO is a natural component of the coupled ocean-atmosphere system and it has a periodicity of about 40-50 days extending to 30-60 days (Madden and Julian 1971, 1994, Zhang 2005).

The MJO is characterized by eastward propagation of areas of enhanced and suppressed tropical rainfall, from the Indian Ocean region towards the very warm waters of the western and central tropical Pacific Ocean. After moving across the cooler waters of the eastern Pacific Ocean, the pattern of anomalous tropical rainfall often reappears over the tropical Atlantic Ocean and Africa. Along with these variations in tropical rainfall, there are distinct patterns of lower and upper level atmospheric circulation anomalies in the global tropics and subtropics. Regions of ascending and descending motion are associated with different phases of the MJO.

The MJO cannot be captured very well by dynamical models because of the problems they continue to have with regard to proper parameterization of tropical convective processes. However, it can be monitored with satellite-derived cloud motion winds and OLR data. The MJO exhibits a high interannual variability, being active in weak La Nina or ENSO-neutral years, and being dormant or absent during strong El Nino episodes. There is evidence that the MJO influences the intensity of El Nino episodes, tropical cyclone activity and the intensity of monsoon systems. The enhanced rainfall phase of the MJO can affect the timing of monsoon onset and the suppressed phase of the MJO can produce monsoon breaks and early withdrawal.

## 4.3 Monsoon Trough

The intense land heating of the northern hemisphere in summer produces a low pressure belt which extends from north Africa to India. The core of this low pressure belt lies over northwest India and adjoining Pakistan and is called the heat low. A deeper heat low is associated with a larger north-south pressure gradient over India and stronger monsoon activity. The monsoon trough is a low pressure area that in its normal position runs from the heat low across the Gangetic plains into the head Bay of Bengal. The oscillations of the monsoon trough during the monsoon season have a close bearing on the rainfall pattern over India. When the monsoon trough is situated at or to the south of its normal position, rainfall activity is strong. When it moves to the north of its normal position, rainfall activity becomes subdued, and in the extreme case of its migrating to the Himalayan foothills there is a break in the monsoon.

The monsoon trough is the only synoptic scale system that contributes to the rainfall activity over India. Monitoring the movement of the monsoon trough is of great help in forecasting the monsoon activity on the short and medium range time scale. In satellite images the presence of cloud clusters helps to identify areas of heavy rainfall. Satellite images clearly bring out the weak phases of the monsoon followed by a revival and buildup. Figure 4.1.1.1 shows the satellite image for 26 May 2006 showing an early and strong monsoon onset over Kerala with an associated onset vortex. There was, however, a midway halt to its progression beyond the peninsula and a weakening of the monsoon current, and on 21 June there was a subsequent revival.

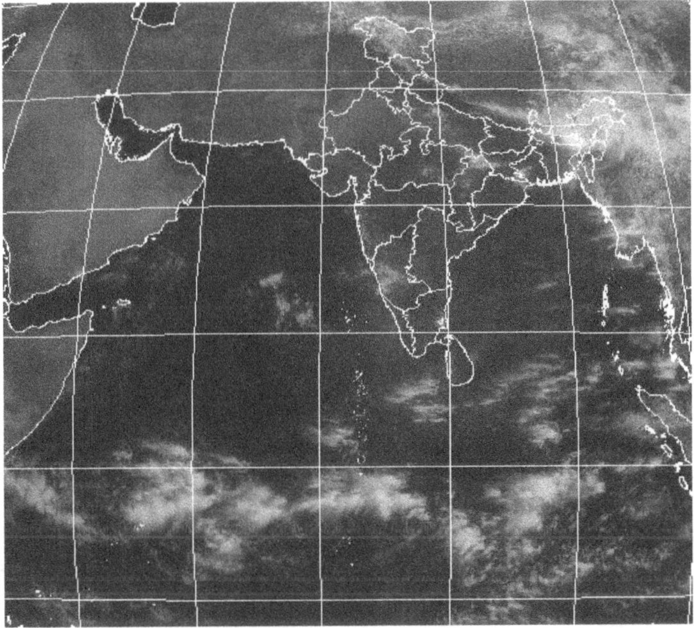

Figure 4.3.1  Kalpana-1 VIS image of 11 June 2006 showing the
reappearance of the SHET in during the weak phase of the monsoon
(Source: IMD)

Satellite pictures have confirmed that the crossequatorial flow over the Indian Ocean has a predominant influence on the monsoon activity. Mishra et al (1991) and Gupta et al (1991) have highlighted the importance of the southern hemisphere equatorial trough (SHET) in regulating the crossequatorial flow thereby leading to the development of the active-break cycle. They found an inverse relationship between the SHET clouding and the monsoon activity over India and persistence of the SHET cloud band beyond the premonsoon months could be a signal of an impending drought. Figure 4.3.1 shows the lack of clouding over the Indian subcontinent and the reappearance of the SHET in the satellite image of 11 June 2006 during the weak phase of the monsoon.

Figure 4.4.2 shows the southwest monsoon in an active phase once again on 3 July 2006. There is a monsoon depression over eastern India, flanked on either side by areas of intense convection on the Arabian Sea and Bay of Bengal, and an off-shore trough along the west coast.

## 4.4 Monsoon Depressions

During the monsoon season, a series of depressions form over the north Bay of Bengal. They generally move in a west-northwesterly direction across the country and often merge into the seasonal low over northwest India. Some of them take a more westerly course and in rare cases even emerge into the Arabian Sea. There are others that take a north-northeasterly course and dissipate over the mountain slopes, causing flooding of rivers which originate there. When depressions linger over central India, it is the peninsular rivers which get flooded instead.

Generally, 4 to 6 monsoon depressions form every year but the number is highly variable. In the 2005 monsoon season, there were 11 low pressure systems, 5 of which intensified into a depression and one into a marginal cyclonic storm. There were other years when only a single depression had formed. The number of monsoon depressions is also seen to have a decreasing long term trend in recent years.

The time interval between two successive depressions is also variable. A depression may follow another in quick succession or after a long interval, ending a prevailing break in the monsoon.

Most of the monsoon depressions have a typical life span of 3 to 5 days, but some may last even up to a week. When the seasonal trough intensifies and a closed low appears, a monsoon depression is likely to grow out of it. The development may occur either in situ or because of the arrival of an easterly wave. In a day or two, the low intensifies into a depression with winds of 22-33 knots. Depressions move at about 4° latitude or longitude per day but if they recurve, they gather speed. In many years, the advance of the monsoon over northeast India takes place in association with a depression.

Unlike tropical cyclones, there is considerable asymmetry in the wind field of a monsoon depression. Winds are stronger to the south of the depression centre, due to the superimposition of the depression perturbation over the westerly flow. The strongest winds are found at the 700 hPa level. The monsoon depression is a system having a cold core below 700 hPa and a warm core aloft. The vertical axis of the depression slopes southwestwards with height.

The southwest quadrant of the depression field has the strongest upward motion and highest rainfall because of three contributing factors: (i) boundary layer frictional effects in areas of cyclonic vorticity which is maximum in the south sector, (ii) warm air advection in the lower troposphere and (iii) vorticity advection in the lower troposphere which contributes to upward motion in the left forward sector and descending motion in the rear sector. Rainfall amounts are generally of the order of 10-20 cm per day and could even be as high as 30 cm per day in isolated cases. Rainfall in the southwest sector of the depression has considerable diurnal variation, with the heaviest falls occurring in the early morning hours.

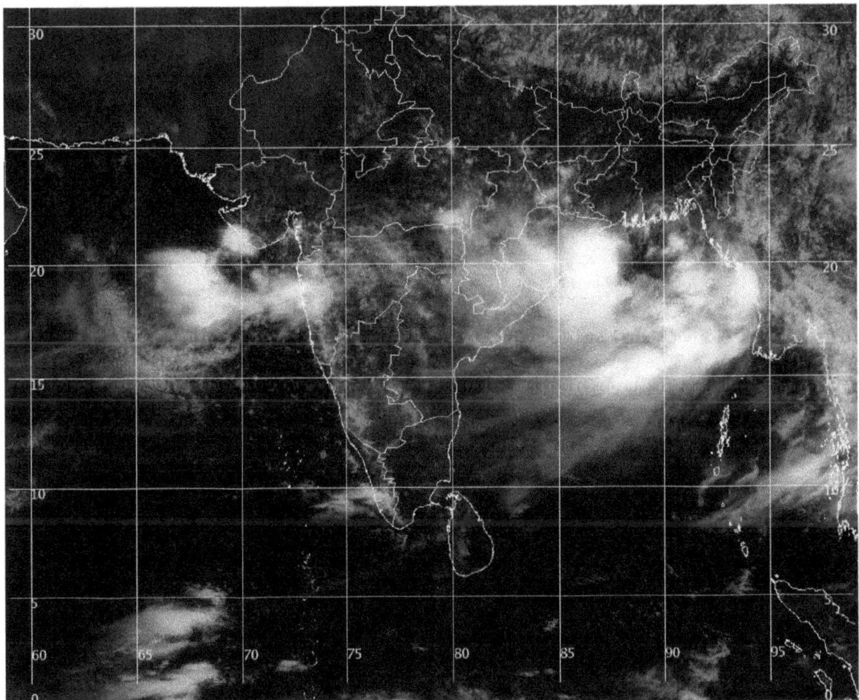

Figure 4.4.1 Kalpana-1 visible channel image of 1 July 2006 showing the formation of a monsoon depression over the Bay of Bengal (Source: IMD)

In its early stages, a monsoon depression induces a low level trough along the Konkan coast, where there is enhanced rainfall activity, which decreases as the depression moves inland. Rainfall activity is high to the west of the depression, but less over southern India and northeast India. When the depression is over central India, Gujarat and Saurashtra get good rainfall because of fresh moisture supply from the Arabian Sea.

A monsoon depression in its formative stage is seen in satellite imagery as a dense amorphous cloud mass over the Bay of Bengal. As it develops further, convective cloud lines tend to get organized and there are pronounced curved cumulus lines at the edge of the dense overcast mass. In the mature stage, the dense overcast gets ragged edges surrounded by curved cumulus lines (Figure 4.4.1 and 4.4.2). Satellite imagery is of great help in detecting the formation of the monsoon depression over the head Bay, in monitoring their movement and intensification. The extent and strength of the cloud organizations in the southwest sector of the depression as revealed in successive images are valuable for giving advance warnings of heavy rainfall to regions lying ahead along the depression's likely track.

Figure 4.4.2 Kalpana-1 visible channel image of 3 July 2006 showing a monsoon depression which has crossed from the Bay of Bengal into Orissa. The curved cumulus lines are seen. The lower portion shows the cirrus blow-off due the strong TEJ winds (Source: IMD)

Characteristic features of monsoon depressions have been described by Srinivasan et al (1971), Chowdhury et al (1983) and Sarkar et al (1988). Chowdhury et al (1985) prepared a mean cloudiness pattern of a monsoon depression by examining 40 case histories during 1977-1981. A moving 20° × 20° lat/long grid was overlaid around the depression centre in the satellite picture and the cloudiness information picked up at each 2.5° grid point which was then averaged to draw isopleths of cloudiness and coefficient of variation. The mean pattern shows a zone of maximum cloudiness 3-4° southwest of the depression centre which could be identified as the principal area of heavy rainfall. The north and southeast extremity of the depression field is an area of suppressed cloudiness. The most distinct feature in all cases is the bright central dense overcast southwest of the depression centre and consisting mainly of stratiform clouds. In about a half of the cases, the centre lay within this amorphous mass, but in the remaining cases the centre was located in the cloudfree zone indicative of a highly sheared environmental flow.

Prasad et al (1990) used 3-hourly INSAT images and other data to construct a composite wind and cloud structure of a monsoon depression. They have also described the cloud features seen in satellite images at different stages of development of the depression. The genesis stage which could be called a well-marked low, is characterized by a long east-west band south of the evolving vortex. In the intermediate stage at which the system becomes a depression, this zonal cloud band reorganizes itself suggesting a rotation of the cloud mass, while cumulus lines appear in the northern sector in the region of low level easterlies and define the centre. The centre could be 1-2° away from the centre fixed on the surface chart. In the intense or deep depression stage, the cumulus lines become more curved and the centre comes closest to the northern edge of the dense overcast. The vertical tilt of the system is the least at this stage. The dense overcast will have a 3-5° diameter with sharp edges and embedded overshooting cloud tops that penetrate the tropopause. The system begins to decay 24 hours later.

Mahajan (2003) and Mahajan et al (2004) have used a variety of satellite data from DMSP SSM/I and Oceansat-1 MSMR and cloud motion winds for studying the development of monsoon disturbances. They found that prior to the formation of a monsoon depression, there is a strengthening of surface winds to the south of the low pressure area over the Bay of Bengal. Highest values of geophysical parameters like integrated water vapour (6-8 gm cm$^{-2}$), integrated cloud liquid water (50-60 mg cm$^{-2}$) and precipitation rate (20-25 mm/hr) and lowest values of OLR (120 w m$^{-2}$) are associated with the southwest sector of a monsoon depression.

Rajeevan et al (2001) examined ten cases of monsoon depressions during 1989-1999 using SSM/I microwave data on daily water vapour, cloud liquid water and rain rate on a 25 × 25 km resolution. However, these data are available only on the ocean. They prepared a composite average pattern for a monsoon depression with respect to a common origin which confirmed the asymmetric distribution of precipitation. While atmospheric water vapour did not show any preferred maxima, there was high cloud liquid water (> 1 mm) in the southwest sector about 200 km from the common centre, which indicated the presence of very deep convective clouds. There the precipitation rate exceeded 12-14 mm/hr, equivalent to 30 cm rain in 24 hours. A secondary maximum was located in the northeast sector with rain rate of 6 mm/hr. Two case studies using TMI data revealed similar features.

## 4.5 Active-Break Cycle

During the southwest monsoon season, rain does not keep falling continuously all over the Indian subcontinent, but there are spells of vigorous monsoon rainfall interspersed by dry spells or periods of weak rainfall activity. This strong intraseasonal oscillation that the monsoon exhibits is also referred to as the active-break cycle.

The southwest monsoon is in its most intense phase during July and August and these are the months of peak rainfall activity over India. However, the monsoon does not maintain a uniform strength all throughout this period and the wet spells are interspersed with dry spells over different parts of the country. At times, when the rainfall becomes very subdued over the country as a whole, the situation is described as a break in the monsoon. Breaks can be of varying intensity and duration, but they have certain common characteristic features. The most prominent synoptic development associated with a break is the shifting of the axis of the seasonal monsoon trough to the foothills of the Himalayas. The large scale atmospheric flow pattern develops many anomalous features. The north-south pressure gradient along the west coast reduces from 9 to 3 hPa and lower tropospheric westerlies weaken along the west coast. On the other hand, westerlies over the Gangetic plains strengthen and their speeds may reach even 50 knots. These winds are dry and transport dust from Rajasthan into the Gangetic plains. The monsoon trough slopes northward instead of sloping southward.

In the case of shorter breaks in the monsoon, only the western end of the monsoon trough may shift to the foothills of the western Himalayas, while the eastern end remains entrenched in its normal position. In this case there is excess rainfall over western Himalayas and reduced rainfall over the plains of northwest India. Such a break may be induced by the passage of a western disturbance. The monsoon activity recovers after the western disturbance has moved away.

When the entire monsoon trough migrates to the Himalayan foothills, there is a prolonged break, which may last for 4 to 5 days or even longer. There is excess rainfall all along the Himalayan foothills, especially on the eastern side, leading to a flooding of the Himalayan rivers, but rainfall is deficient over large parts of the country.

While July and August are the months of peak monsoon rainfall, breaks are also most common in these months. If a break-like situation develops in June, it is not called a break, as the monsoon has not firmly established itself by then all over the country. Likewise, in September, a weakening of the monsoon is not called a break, as it may lead to an early withdrawal of the monsoon and it may not revive after the break.

As the breaks in the monsoon strongly modulate the All-India Summer Monsoon Rainfall (AISMR), they have been the subject of several studies, in which different definitions of a break are also to be found (Gadgil et al 2003, De et al 2002, Webster et al 1998). In a very recent paper on this subject, Rajeevan et al (2006) have made an analysis of the active-break spells of the monsoon using the newly available high-resolution daily gridded rainfall data of the Indian region. Break (active) periods during the monsoon season were identified as those in which the standardized daily rainfall anomaly averaged over the Central Indian region 21-27° N, 72-85° E, is less than -1.0 (more than 1.0). They did not find any evidence of any statistically significant trends in the number of break or active days during the period 1951-2003.

The most prolonged break was in July 2002 when AISMR was deficient by as much as 19 %. There were long breaks during August 2005 and rainfall was deficit by 28 % for the country as a whole in that month. This has happened only twice before in 1901 and 1920.

In most cases, there are no clear advance synoptic indications of an impending break in the monsoon or its cessation, but satellite images provide some signals. A break may be caused by a monsoon low or depression moving from the head Bay towards Assam. The monsoon usually recovers gradually to a normal pattern. The cessation of the break may sometimes be induced by the formation of a monsoon depression in the Bay or the passage of a low pressure wave across south Indian peninsula that builds up the pressure gradient along the west coast. Satellite imagery is of great value in monitoring the migration of the monsoon trough from its normal position to the foothills and back. Satellite images during break situations are characterized by a conspicuous absence of clouds over a large part of the country.

In a recent work (Jenamani 2004), two intermediate phases, one in which the monsoon trough is migrating from its normal position to the foothills and another in which the monsoon trough is returning from the foothills to its normal position have been defined in detail with respect to rainfall, pressure and circulation features. He has concluded that in the pre-break phase, there is large negative pressure anomaly and excess rainfall over north India and large positive pressure anomaly and deficient rainfall over peninsular India. In the post-break revival phase, the opposite conditions prevail. He also found that in the pre-break phase, monsoon disturbances form over extreme north Bay of Bengal and in the revival phase, they form over south Bay of Bengal. If these disturbances move northward or northwestward the monsoon revives, but if they move westward the break only gets prolonged and intensified.

Goswami et al (2003) formulated an empirical technique that had a useful skill in predicting the monsoon breaks up to 18 days in advance and the return to active conditions up to 10 days in advance. Dwivedi et al (2006) have enlarged the scope of this technique by using the theory of regime transitions in a forced Lorenz model. They showed that the peak anomaly in an active regime can be used as a predictor for the duration of the following break.

Xavier and Goswami (2007) have proposed a physically based empirical real time forecasting strategy to predict the subseasonal variations of the Indian summer monsoon up to 4-5 pentads (20-25 days) in advance. The method makes use of analogues in the properties of monsoon intraseasonal oscillations (ISOs) as revealed by OLR patterns. Skillful predictions can be made up to 5 pentads when started from an active initial state, whereas the limit of useful predictions is reduced to 2 or 3 pentads when started from break initial conditions.

## 4.6 Withdrawal of Monsoon

Although the period 1 June to 30 September is commonly regarded as the southwest monsoon season, it is more so because of the convenience of generating climatological statistics of rainfall and other atmospheric parameters than for sound physical reasons. While the onset of monsoon over Kerala occurs around 1 June, northwest India may not get its monsoon rains until the end of June or even much later until the middle of July. However, all premonsoon rainfall from 1 June till the actual commencement of the monsoon rains, gets clubbed with the AISMR for climatological purposes. Likewise, the southwest monsoon starts losing its strength in early September over northwest India, where the monsoon season is effectively of only two months duration. On the other hand, over large parts of central and southern India, rains persist much beyond the date of 30 September, which are not accounted in the AISMR.

The withdrawal of the southwest monsoon from northwest India is an event that is in no manner comparable to its onset over Kerala. While the onset over Kerala has been a subject of great scientific interest, there have been very few studies of the withdrawal of the monsoon. The withdrawal process commences with a southward migration of the 200 hPa ridge position, establishment of low level (850 hPa) anticyclonic flow over northwest India, a reduction in the atmospheric moisture content and of course the cessation of rainfall. This is the beginning of the transition to the winter circulation pattern, and the return of the ITCZ to its southerly position. There is no clear definition of the withdrawal, however, and there is no rule as to how long to wait before declaring that the monsoon rains have come to an end. There are many instances when the monsoon withdrawal has been followed almost immediately by rains in north India as a result of the passage of a western disturbance or over peninsular India because of strong thunderstorm activity.

Generally, by 15 October, the southwest monsoon withdraws from the country (Figure 1.4.2) except from the southern peninsula, where the northeast monsoon continues to bring more rains as described in the next section.

### 4.6.1 Forecasting the Dates of Withdrawal

Not many studies have been made of the withdrawal of the monsoon, primarily because of its lesser practical importance and also because the return of the southwest monsoon is linked with the setting in of the northeast monsoon over southern parts of India. In recent times, there has been a tendency to treat both the onset and withdrawal of the monsoon at par, as the forward and inverse aspects of the same physical process (Fasullo et al 2003, Goswami et al 2005 and Xavier et al 2007). The basic premise of Goswami and Xavier is that the length of the rainy season is controlled by the sign reversal of the meridional gradient of tropospheric (200-700 hPa) temperature which in turn is influenced by ENSO. The atmosphere responds to the diabatic heating associated with the ENSO SST in spite of the distance between the monsoon region and the core convection associated with ENSO, with the tropical wave dynamics providing an atmospheric bridge. During El

Nino years, the onset of the monsoon gets delayed by 9 days on an average while the withdrawal gets advanced.


## 4.6.2 Northeast Monsoon

Even after the withdrawal of the southwest monsoon, rainfall continues until mid-December over southern parts of India, and up to the first week of January over Sri Lanka. This rainfall is produced by the northeasterly winds in the lower troposphere associated with the northeast monsoon. Compared with the southwest monsoon, the northeast monsoon is a phenomenon of a much smaller scale, and it affects only the southern parts of the Indian peninsula during the months of October to December.

The mean date of onset of the northeast monsoon over Tamil Nadu is 20 October with a standard deviation of a week. However, there have been years of very early onset (5 October 1943) as well as extremely delayed onset (11 November 1915). The onset of the northeast monsoon is not rigidly defined, but it is known to be associated with several distinguishing features like the withdrawal of southwest monsoon up to coastal Andhra Pradesh, persistence of easterly/northeasterly winds up to 850 hPa in the lower troposphere and the occurrence of fairly widespread rainfall over coastal Tamil Nadu and adjoining areas. The onset is also marked by a reversal of north-south pressure gradient with the lowest pressure being over the extreme southern peninsula, strengthening of upper tropospheric westerlies over north India and increase in moisture content over the southern peninsula.

The northeast monsoon gets intensified with the passage of easterly waves and low pressure areas across the Bay of Bengal, some of which may grow into tropical cyclones. As the northeast monsoon season advances, the cyclones move at relatively lower latitudes. In November and December, cyclones have a tendency to move towards Tamil Nadu.

For Tamil Nadu, northeast monsoon rainfall is more plentiful than southeast monsoon rainfall. The annual normal rainfall of Tamil Nadu is 100 cm, out of which 35 cm is received during the southwest monsoon season and 48 cm in the northeast monsoon season. The northeast monsoon rainfall has a standard deviation of 14 cm making it highly variable.

The northeast monsoon season ends in December. Again there is no specific definition of the withdrawal of the northeast monsoon. After December, the near-equatorial trough moves further south and the formation of tropical cyclones near the equator is inhibited.

Khole et al (2003) have composited the area-weighted rainfall for the five meteorological subdivisions of India forming the core region of the northeast monsoon and found that it is 11 % higher than the long period average in El Nino years and 6 % lower in La Nina years.

Raj (1996) has studied the thermodynamical aspects such as precipitable water vapour, upper air temperatures and moisture fluxes during different phases of the northeast monsoon. Suresh et al (2001) made a similar analysis based upon TOVS-derived outgoing longwave radiation (OLR) and precipitable water vapour for 1996-1998 over the northeast monsoon domain. They found that in the active phase of the northeast monsoon, the lower and upper atmosphere is warmer and the middle atmosphere cooler than in the weak phase. The OLR distribution suggests that during the active phase, the rainfall is maximum over south coastal Tamil Nadu and decreases towards the Bay of Bengal, while in the weak and postwithdrawal phases, there is higher rainfall over the Bay than over Tamil Nadu. Suresh et al (2002) have also made a study based on TOVS data of atmospheric boundary layer parameters during the northeast monsoon.

## 4.7 Flood Forecasting

With the rivers of central and peninsular regions of the country, flooding is an annual feature, caused by heavy monsoon rainfall over the catchment areas. Flooding in the Himalayan rivers is caused by heavy precipitation in the upper catchments and is aggravated by factors such as rivers changing their course, increase in the silt load, construction of embankments, etc. There is a well-organised system in this country for forecasting of river floods, which is run by the Central Water Commission with the active involvement of the Flood Meteorological Offices of the India Meteorological Department.

The major Indian rivers are those of the Himalayan Region: Brahmputra River (North-eastern States and West Bengal), Ganga River (Bihar and Uttar Pradesh) and Northwest Rivers (Jammu and Kashmir, and Punjab). The rivers of central and peninsular India are West-flowing Rivers (Gujarat) and East-flowing Rivers (Andhra Pradesh and Orissa).

With Himalayan rivers, flooding is an annual feature. When there is heavy precipitation over the Himalayas, river waters gush down because of steep slopes causing swollen tributaries and spilling over banks downstream. These rivers have a tendency to change course and form channels. Floods spread over much wider areas and silt load aggravates the situation.

The rivers of central and peninsular India are numerous and they flow in opposite directions. Narmada and Tapi flow into the Arabian Sea and Mahanadi, Godavari and Krishna into the Bay of Bengal. The problem of flooding with these rivers is less acute than with Himalayan rivers. They have relatively well-defined courses. It is the heavy monsoon rainfall that causes floods. Delta regions are more floodprone due to congestion.

River floods can be predicted because there is considerable time lag between the occurrence of heavy rainfall in the upper catchment and the consequent buildup of the flood flow in the river, and its travel to a downstream area. Such a lead time is not available in case of drainage congestion caused by local rainfall. Also, the propagation of a flood wave in a river channel is easier to compute. Mathematical or physical modelling of city drainage is, from a hydraulics point of view, a far more complex problem. It also parallelly needs a quantitative prediction of the rainfall amount and rate on a scale that will match the scale of the hydraulic model. As of today, the state of art in these areas is rather primitive.

A good flood forecasting model is one that is able to forecast parameters such as the timing of the flood, its duration, depth and discharge. Hydrological process models require meteorological and hydrological inputs and yield downstream flow as their output. Mathematical hydraulic models need a full description of the river channel but are able to forecast many characteristics of the flood. There are some models which are a kind of combination of the two types.

The hydrological data requirements include stream-flow measurements as well as peak flow measurements. Meteorological data requirements pertain not only to intensity and duration of rainfall that has occurred, but precipitation forecasts for the next 24-48 hours and past data for rainfall-runoff relationships. Models also need topographic data to delineate the possible extent of inundation and the watershed area, geological and soil data, and demographic data.

## 4.7.1 Heavy Rainfall Events

On the time scale of a day, there are many past instances of Indian stations having recorded as much as half of their annual rainfall, or even more than their annual rainfall, in one single day (Dhar et al 1981). Rainfall of 50 cm or more in a 24-hour period is not an uncommon phenomenon at all (Rakhecha et al 1980).

The observatory at Santa Cruz in north Mumbai recorded a rainfall of 94.4 cm during the 24 hours ending at 8:30 am on 27 July 2005, while the Colaba observatory in Mumbai's southern tip recorded barely 7.3 cm in the same period. Rainfall over Vihar lake was 105 cm, even higher than Santa Cruz. The previous record of heaviest 24-hour rainfall over Mumbai was 58 cm for Santa Cruz and 37 cm for Colaba on 5 July 1974 (IMD 2005). Comparatively speaking, only Santa Cruz broke the previous record, but for Colaba the rainfall was in no way unusual.

Heavy rainfall (more than 20 cm) is quite common for Mumbai during the onset phase of the monsoon. It is caused by a convergence of the dry winds blowing at that time from the north with the advancing moist southwesterly winds of the monsoon, coupled with the development of an onset vortex either over the Arabian Sea or the Bay of Bengal. However, after the monsoon has set in and goes into its active phase, the synoptic situation is conducive to the occurrence of very heavy rains over Mumbai when it has the following features collectively: (1) development of a low pressure area over the northwest Bay of Bengal, (2) intensification of the monsoon trough and development of embedded convective vortices over central India, (3) strengthening of the Arabian Sea current of the monsoon, and (4) superpositioning of a mesoscale off-shore vortex over northeast Arabian Sea and its northward movement. All these conditions were met on 26 July 2005 (Shyamala, 2005). The Mumbai downpour was the result of a combination of synoptic scale weather systems which span across 1000-2000 km, with mesoscale systems which are localized and extend over 20-30 km only.

## 4.7.2 Factors Responsible for Flooding

Floods are caused by a variety of factors, meteorological, hydrological and anthropogenic. The physical characteristics of floods vary greatly depending upon the areal extent of inundation, the duration of flooding, the suddenness of onset. The characteristics of flooding differs greatly in rural and urban settings. Flash floods are common in the mountain areas of India and are caused by sudden heavy rainfall or cloudbursts. They take the downstream populations by surprise as they are not even aware of the heavy rain which has occurred somewhere else. Widespread storms like monsoon depressions cause flooding on a regional scale. The storm surge associated with tropical cyclones leads to coastal flooding.

The magnitude of the flooding is dependent on hydrological conditions like the existing soil moisture, the level of shallow ground water and the surface infiltration rate. Among the long term factors that play a role are those such as deforestation, erosion, sedimentation and changing land use patterns.

## 4.8 Statistical Models for Subseasonal Rainfall Prediction

As per Webster et al (2004), convection over the eastern Indian Ocean is stronger than that over the western Indian Ocean but lags behind it by a few days, and it is out of phase with convection over India. The northward extension of convection occurs primarily in the eastern Indian Ocean and coincides with the active periods of the monsoon. They used this knowledge for the prediction of intraseasonal oscillations in the south Asian monsoon region using a new type of statistical model. This is a dynamically based Bayesian statistical scheme that combines wavelet analysis and linear regression, which they have called the Bayesian wavelet-banding method, following a similar scheme used earlier for hydrological forecasting in Colombia. Webster et al have discussed their forecasts in the 15-30 day time range for precipitation over central India and the smaller regions of Orissa and Rajasthan, in addition to Ganges and Brahmaputra River discharge into Bangladesh.

Webster et al chose 10 predictors which are physical components of the monsoon intraseasonal oscillations. They included OLR over equatorial Indian Ocean and central India, wind over the Arabian Sea and equatorial Indian Ocean, soil moisture and surface pressure over central India, the strength of the Somali jet and tropical easterly jet. SST did not figure in their list of predictors. These predictors had a certain pattern of behaviour, in terms of magnitude and tendency, several pentads prior to the occurrence of an high precipitation event over central India. The problem then reduces to building a statistical prediction scheme that takes into account the conditional behavior of the entire set of predictors. Thus, each future point is independently forecast with a regression model constructed for each period to be predicted.

Webster et al's 20-day forecasts could discern the phase of the major variations but the rainfall amounts were generally underestimated. In addition to forecasting intraseasonal variability of the monsoon, the scheme was also able to determine the commencement of the monsoon rains over central India with some accuracy.


## 4.9 Dynamical Models

After computers like the IBM-1620 and CDC-3600 came to India in the 1960s, a new avenue was opened to Indian meteorologists for constructing numerical models of the atmosphere. One of the earliest numerical simulations of the monsoon was made by Murakami, Godbole and Kelkar (1970) at the Institute of Tropical Meteorology, Pune. They simulated a near-steady state monsoon in a zonally symmetric framework along the longitude 80° E in the month of July. They designed a primitive equation model with a $\sigma$-system in which the atmosphere was divided into 8 layers. The top of the frictional boundary layer was taken as level 8½. In the horizontal only $y$-variation was considered, with 18 grid points at 5° latitude interval. In the $x$-direction, symmetry was assumed.

Radiation heating and cooling was computed within the model according to the scheme described by Godbole, Kelkar and Murakami (1970), using the distribution of absorbing gases and clouds for the month of July. During the iterations, whenever the vertical lapse rate of temperature exceeded the wet (dry) adiabatic value with saturated conditions, it was adjusted to the wet (dry) adiabatic value by a redistribution of static energy. SST was kept at a constant value of 300 °K, while the land surface temperature was computed from the heat balance of the surface fluxes.

The model was run with a calm initial state and the initial vertical distribution of temperature was prescribed as per the standard atmosphere. The model was run with a 10-minute time step for 80 days and experiments were repeated without the Himalayan orography.

Murakami, Godbole and Kelkar (1970) were successful in simulating some of the large scale features of the monsoon circulation and the wind speeds were of the same order as those in climatological normals. Their result of particular importance was that when the Himalayan mountains were removed from the orographic profile along the longitude 80° E, the zonal circulation became much weaker, with lower level westerly winds of 10 knots and upper level easterly winds of 20 knots only. Thus a realistic parameterization of the Himalayan mountains is crucial to the numerical simulation of the monsoon.

Subsequent to the pioneering work of Murakami et al, there have been several modelling studies of the monsoon that have been able to simulate reasonably well many features of the monsoon such as the tropical easterly jet, the Somali jet and the low level westerly flow over India. However, even as of today, the most challenging problem is that of simulating realistically the mean rainfall pattern of the monsoon.

It is a discouraging fact that there has been no effort on the part of Indian meteorologists to build an Indian model in spite of the continuous upgradation of computational resources and the growth of institutional infrastructure. Both NCMRWF and IMD have made use of foreign models for operational purposes. The approach has all along been to test foreign models over Indian conditions and judge their usefulness for predicting the monsoon. The search for an ideal model that could be used for monsoon prediction is still on.


## 4.10 Atmospheric General Circulation Models

The National Centre for Medium Range Weather Forecasting (NCMRWF) was established at New Delhi in 1989 for making medium range forecasts up to 10 days using Atmospheric General Circulation Models (AGCMs). The global T80 model was able to provide a signature of the large scale weather systems fairly well particularly in the winter season, but not quite on a regional scale. In order to make accurate forecasts of the mesoscale systems such as the western disturbances, severe thunderstorms, tropical cyclones and heavy rainfall episodes during the active monsoon season, high resolution mesoscale models such as MM5 and ETA were also being run at NCMRWF on real time basis.

On 1 January 2007, NCMRWF implemented a new Global Forecast System (GFS) at T254L64 resolution on the IBM P5 cluster-based Param Padma and Cray-X1E computer systems (Rajagopal et al 2007). The new GFS and the data assimilation system has been adopted from NCEP. The horizontal representation of model variables is in spectral form with transformation to a Gaussian grid for calculation of nonlinear quantities and physics. The quadratic T254 Gaussian grid has 768 grid points in the zonal direction and 384 grid points in the meridional direction. This corresponds to a horizontal resolution of about $0.5° \times 0.5°$ latitude/longitude. In the vertical there are 64 $\sigma$-levels, out of which 15 are below 800 hPa and 24 are above 100 hPa. The time step for model integration has been set at 7.5 minutes. The parameterized model physical processes are: gravity wave drag and realistic orography, radiation, cumulus convection (Simplified Arakawa Schubert scheme), shallow convection, large scale condensation, diagnostic clouds, PBL, air-sea interaction and land-surface processes.

In the current implementation only conventional data sets and satellite cloud motion vectors are being assimilated in the GFS. One analysis cycle and a 7-day forecast run takes about 3 hours on Cray X1E computer. Weather analysis and predictions generated by the system are reasonably good and match well with those produced by other leading NWP centres.

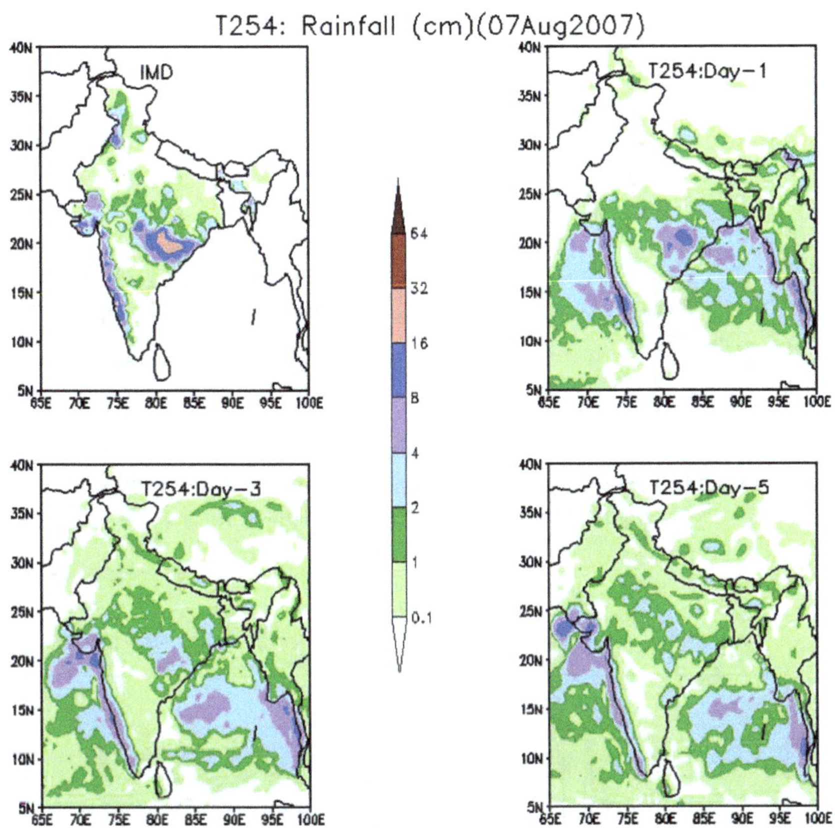


Figure 4.10.1 Comparison of observed rainfall distribution for 7 August 2007 with NCMRWF T254 model predictions for Day-1, Day-3 and Day-5 (Source: NCMRWF)

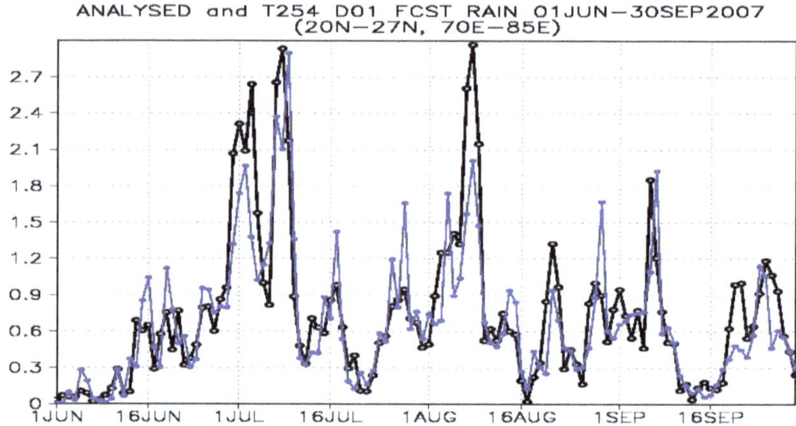

Figure 4.10.2 NCMRWF T254 model day-1 predictions of the daily
monsoon rainfall averaged over central and northwest India (blue line)
compared with observed rainfall (black line)
(Source: NCMRWF)

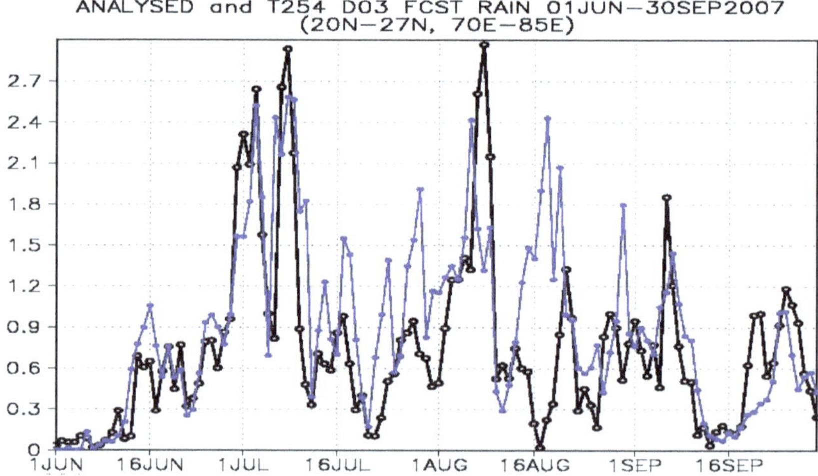

Figure 4.10.3 NCMRWF T254 model day-3 predictions of the daily
monsoon rainfall averaged over central and northwest India (blue line)
compared with observed rainfall (black line)
(Source: NCMRWF)

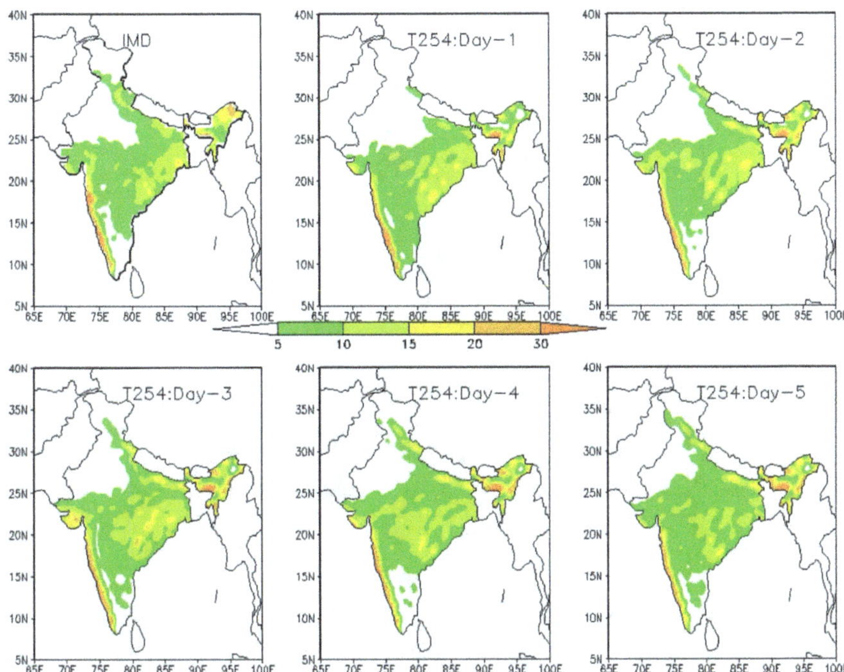

Figure 4.10.4 NCMRWF T254 model day-1 to day-5 predictions of the spatial distribution of daily monsoon rainfall (mm/day) averaged individually over JJAS 2007 compared with observed rainfall (Source: NCMRWF)

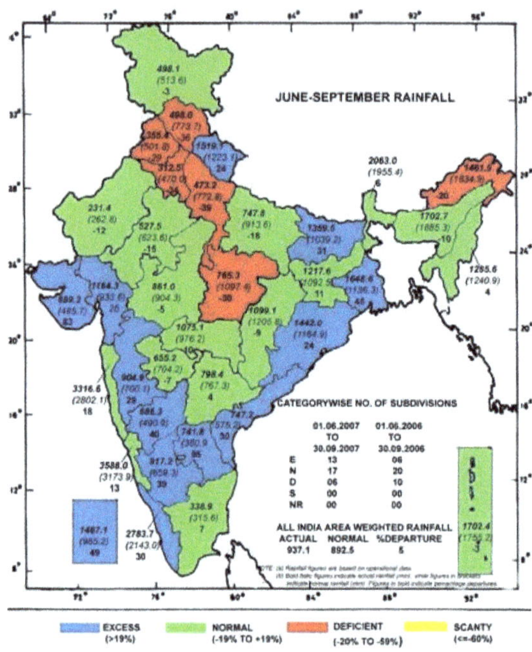



Figure 4.10.5 Subdivisional rainfall departures from normal during the monsoon season of 2007 (Source: IMD)

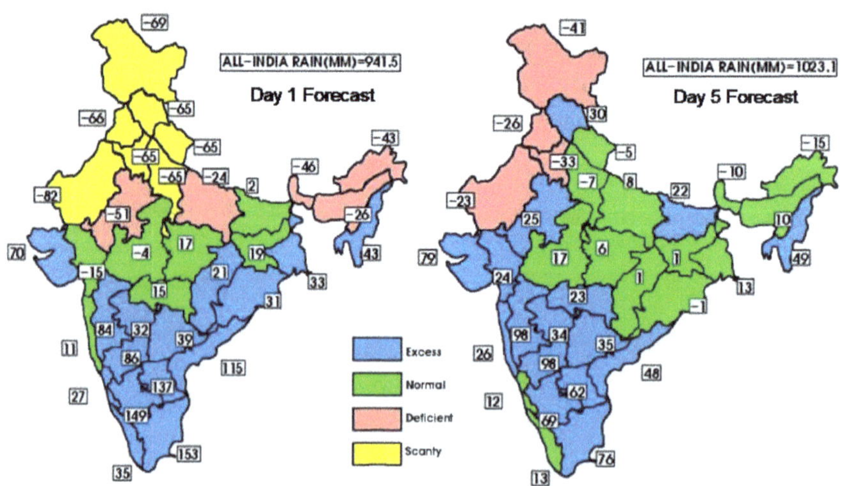



Figure 4.10.6 NCMRWF T254 model day-1 and day-5 predictions expressed as subdivisional rainfall departures from normal (Source: NCMRWF)

The performance of the NCMRWF T254 model in a case study of a monsoon depression is shown in Figure 4.10.1. This depression had formed over west central Bay of Bengal off West Bengal-Orissa coast on 5 August 2007. It concentrated into a deep depression, crossed the coast the next day and moved northwestwards. It continued to move across the central parts of the country and weakened into a low pressure area over Gujarat on 9 August. There was good rainfall activity over most of the country in association with this system and this was captured well by the day-5, day-3 and particularly day-1 forecasts for 7 August 2007 by the T254 model. The predicted region of heavy rain and the location of the core area of rainfall activity to the southwest of the depression field, compare well with the analysis.

A necessary test of a model is to verify how well it can predict the observed intraseasonal and spatial variability of the monsoon rainfall. Figures 4.10.2 and 4.10.3 show a comparison of the NCMRWF T254 model day-1 and day-3 predictions of the daily monsoon rainfall averaged over central and northwest India with the observed rainfall for the period 1 June to 30 September 2007. The day-1 predictions compare fairly well with the observed rainfall values, and there is almost a one-to-one correspondence between the rainfall peaks with regard to their days of occurrence. In other words, no spurious maxima and minima were predicted by the T254 model for day-1. However, some of the smaller rainfall peaks of the series are seen to be overestimated while the higher rainfall amounts have been generally underestimated. In the case of day-3 forecasts, however, the comparison is not so good and there are many instances of a gross overestimation of rainfall.

If the day-1 to day-5 rainfall forecasts are individually averaged over the four months of the monsoon season 2007 and compared with the observed rainfall, the patterns show a general resemblance (Figure 4.10.4), but there are many finer differences between the observed and predicted rainfall amounts and among the five different predicted patterns themselves.

IMD follows the practice of producing a rainfall map at the end of the monsoon season that gives the rainfall departures over different subdivisions in four categories: excess, normal, deficient and scanty. In the monsoon season of 2007, 13 subdivisions received excess normal, 17 normal and 6 deficient. Five out of these 6 deficient subdivisions formed a north-south strip running from Himachal Pradesh to east Madhya Pradesh while the excess subdivisions formed a V-shaped pattern covering the eastern and western parts of the peninsula and adjoining areas (Figure 4.10.5). The NCMRWF day-1 and day-5 predictions, when presented in the subdivisional pattern of IMD, agree well with the IMD categories over southern, western and eastern India. However, there are some obvious differences over northern, northwest and northeast India. The predicted patterns also differ among themselves over these parts of the country (Figure 4.10.6).

**4.11 Regional, Limited Area and Mesoscale Models**

IMD has been running operationally the Limited Area Model (LAM) of Florida State University for short-range numerical weather prediction over the monsoon region for more than 20 years. This model has a $1° \times 1°$ lat./long. resolution and 12 vertical sigma levels. The boundary conditions and the initial analysis are taken from the operational version of NCMRWF global spectral model. The flow field and precipitation from real time short-range forecasts associated with the Indian region is well-represented by this limited area model (Roy Bhowmik et al 2001). Because of its low resolution, the orographic rainfall associated with the Western Ghats is under-estimated during the monsoon, but at other times, when the orographic rainfall is small, the skill of the model is found to be much higher. By prescribing a realistic initial moisture field from INSAT infrared channel data over the Bay of Bengal and Arabian Sea, the model skill of the precipitation forecast associated with movements of monsoon depression was shown to improve considerably (Rama Rao et al 2001). When the model resolution was increased to 50 km in the horizontal and to 16 levels in the vertical, it could capture the mesoscale convective organization associated with cyclonic storms and monsoon depressions more realistically as well as the heavy rainfall belt along the Western Ghats (Roy Bhowmik 2003).

The Pennsylvania State University and the U. S. National Center for Atmospheric Research (NCAR) pioneered the development of a numerical mesoscale model in the 1970s. Their model went through a continuous improvement with the involvement of several institutions as a community exercise. The fifth generation version, popularly known as MM5, along with auxiliary processing programs is freely provided and supported by NCAR for worldwide use (MM5 2008). MM5 is a limited-area terrain-following sigma-coordinate model designed to simulate and predict mesoscale atmospheric circulation. It has several features like non-hydrostatic dynamics and several physics options, and capabilities for multiple-nesting, multitasking and four-dimensional data-assimilation. MM5 is a regional model, and the initial conditions and the lateral boundary conditions required by it have to be obtained from another model running over a larger area.

Besides MM5, many other mesoscale models have been parallelly developed, such as the Regional Atmospheric Modeling System (RAMS) of Colorado State University, the Advanced Regional Prediction System (ARPS) of the

Center for Analysis and Prediction of Storms at University of Oklahoma and the mesoscale step-mountain ETA ($\eta$-coordinate) model of the U. S. National Centers for Environmental Prediction (NCEP) for research as well as operational forecasting. The MM5, RAMS, ARPS and ETA models have been tested in India and several case studies of western disturbances, mountain weather systems, active monsoon situations and tropical cyclones, which have shown that mesoscale models are capable bringing out the finer features of the rainfall distribution much more realistically than global models (Das 2002, Mohanty et al 2003, Vaidya 2006, 2007).

Das (2002) demonstrated that the MM5 model had a good capability of making forecasts up to 72 hours in advance over the Indian region. He used the MM5 model with triple nested domains at 90, 30 and 10 km resolutions, with the initial and lateral boundary conditions derived from the operational global models of NCMRWF. The data assimilation was also done indirectly through the global models. During 13-17 August 2001, there was intense monsoon activity over many parts of India. Gujarat, Konkan, Goa, coastal Karnataka, Kerala and the Western Ghats received very high rainfall. The Delhi region received rainfall of the order of 19 cm on 12-14 August. In two separate studies, Das (2003, 2005) showed that 48 hr MM5 model forecasts correctly indicated such high rainfall amounts, particularly when the model was run in the cloud-resolving scale with an explicit cloud microphysics scheme.

In an investigation of similar situations, Mohanty et al (2003) have found that the MM5 model was useful in simulating the track of the Orissa supercyclone of 29 October 1999 and the associated rainfall, the heavy rainfall event over West Bengal during 18-21 September 2000, and an intense western disturbance of 21-25 January 1999.

Mesoscale models are undergoing a continuous development and a new and higher version mesoscale model called the Weather Research and Forecasting Model (WRF) is currently in operational use at NCEP. The WRF model is the outcome of a collaborative partnership among many U. S. agencies and universities (WRF, 2008). It features multiple dynamical cores, a 3-dimensional variational (3DVAR) data assimilation system, and a software architecture allowing for computational parallelism and system extensibility. WRF can serve a broad spectrum of applications across different spatial scales. It allows researchers to conduct simulations reflecting either real data or idealized configurations. It is flexible and computationally efficient, while offering the advances in physics, numerics, and data assimilation contributed by the research community.

Both the MM5 and WRF models are being run on a real time test mode basis in India at IMD and NCMRWF, and the MM5 model has been providing semi-operational forecasts for the mountain regions of the western Himalayas since 2002. The Centre for Development of Advanced Computing, Pune, is also involved in real time forecasting experiments using the WRF model (Kesarkar et al 2005).

In a study by Rama Rao et al (2005), the WRF and ETA models have been used to simulate some recent heavy rainfall events that occurred over Maharashtra and coastal Tamil Nadu and Andhra Pradesh, and a monsoon depression / tropical cyclone that developed during the withdrawal phase of the southwest monsoon in 2004. They have also performed sensitivity experiments with the WRF model to test the impact of microphysical and cumulus parameterization schemes in capturing the extreme weather events. Their results have shown that the WRF model, with the microphysical process and cumulus parameterization schemes of Ferrier and Betts-Miller-Janjic, was able to capture the heavy rainfall events better than the other schemes. The WRF model was able to predict mesoscale rainfall more realistically in comparison to the Eta model of the same resolution.

Vaidya (2006) used the ARPS mesoscale model with two alternate cumulus parameterization schemes of Kain-Fritsch and Betts-Miller-Janjic for two case studies for the monsoon season. One was of a low pressure area that had formed off the north Andhra-south Orissa coast on 28 July 1998 and had later moved inland. The other was of a low pressure area that had formed over central Bay of Bengal on 13 June 1998 and developed into a monsoon depression.

## 4.12 Using AGCMs for Seasonal Monsoon Rainfall Prediction

In recent years, because of the availability of computer power that was unimaginable a decade earlier, it is becoming possible to design models of increasing complexity. However, our knowledge of many of the physical atmospheric and oceanic processes is yet far from complete, while some of the processes like cloud formation and rain, which are otherwise known well, have not yet been parameterised in models in a satisfactory manner. Furthermore, while the models may be capable of assimilating a variety of complex inputs, in reality such data may not necessarily be available either in the desired accuracy or in terms of coverage. All these factors lead to a situation wherein no two models behave alike and the results are model-dependent, leaving a lot to the users to ponder about before accepting the results and putting them to use.

In the particular case of the monsoon, the anomalies in the surface boundary conditions like snow cover or SST have been used for over a century in statistical models to predict the monsoon rainfall. However, it must be accepted that the complete chain of physical and dynamical connections has not been identified even today. The present generation of AGCMs have such large systematic errors in simulating both the mean and the variability of monsoon rainfall, that it is difficult to isolate the result of boundary forcings over the system noise produced by the intrinsic variability due to internal dynamics.

The Atmospheric Model Intercomparison Project (AMIP), mooted in 1990 (Gates et al 1992), was the first organized international effort to determine the systematic errors of AGCMs on seasonal and interannual time scales. For purposes of comparison and validation, all models were run for an identical period of ten years 1979-1988 and with observed SST and sea ice conditions. This was followed by several diagnostic subprojects with emphasis on specific physical processes and phenomena.

Under the AMIP diagnostic subproject on the monsoons, an analysis of the seasonal precipitation associated with the African, Asian (Kang et al 2002) and the Australian-Indonesian monsoon was carried out. Gadgil et al (1998) have discussed the interannual variation of the Indian monsoon simulated by 30 AGCMs in this major international effort. The period chosen for the AMIP runs 1979-1988 included two major ENSO events of 1982-83 and 1987-88 and covered a large range of variations of the Indian monsoon rainfall. The AMIP monsoon subproject was therefore able to throw light on how well the models existing at that time could simulate the monsoon variability on the seasonal to interannual time scale.

Gadgil et al (1998) showed that the seasonal migration of the major rainbelt observed over the African region, was reasonably well-simulated by almost all the models. However, the seasonal variation of the rainbelt of the Asia-West Pacific sector was captured well only by some models, termed as class I models, but not by others, termed as class II models. For the Indian monsoon, a further complexity arises as during July-August there is a primary rainbelt over India, a secondary one over the equatorial Indian Ocean and a third one over the Himalayan foothills. Whereas in some class I models, e.g. GFDL, the rainfall over the Indian monsoon zone occurred in association with the seasonal migration of the planetary scale rainbelt, in some class II models, e.g. COLA, the rainfall was associated with a regional system with a smaller longitudinal extent. Very few models were able to distinguish well between good and poor monsoon seasons, but class I models showed better skill in this respect. The mean rainfall pattern for the models of class I over the Indian region was closer to the observed one than for class

II. The simulation of the mean rainfall pattern by NCEP2 appeared to be the most realistic. Some models were able to capture only the deficient rainfall years like 1979, 1982, 1986 and 1987 and some other models could simulate only the years of excess rainfall.

Gadgil et al (2002) have discussed the early results of the AMIP-2 project in which 20 different models were compared for their ability to simulate the monsoon features in the contrasting years 1979, 1982, 1983, 1987 and 1988. It was observed that the NCEP model was able to get the correct sign of the departure from the mean during all the five years, the ECMWF model in four out of five years, and the COLA model in only two years. The magnitude of the deficit or excess was not realistically simulated by any of the models, with NCEP overestimating it in three out of five years and ECMWF underestimating it in four years.

The reason ascribed by Gadgil et al to the behaviour of these models is that in comparison with the rest of the tropics, over the Indian longitudes there are two favourable zones for the rainbelt to occur: one over the heated subcontinent and another over the warm waters of the equatorial Indian ocean. The models tend to simulate a rainbelt that gets locked into either of these locations, while in reality the rainfall activity keeps on oscillating between them. The summary of the AMIP findings was that no single model out of the selected foreign models available at that time could be said to be able to correctly predict all the seasonal and intraseasonal elements of the monsoon that are of importance.

Another model intercomparison project had been organized under the CLIVAR Monsoon Intercomparison Project in which 11 AGCMs had been tested. They did not show any skill in their ensemble simulations of the summer rainfall anomalies during the 1997-1998 El Nino event. Wang et al (2004) attributed this failure to the neglect of the air-sea interaction processes in these models. In a later work, Wang et al (2005) extended their analysis to test the simulation skill of five state-of-the-art AGCMs in seasonal precipitation for a 20-year period 1979-1998. These models were forced by identical observed SST and sea-ice, following the design of the Atmospheric Model Intercomparison Project (AMIP). Each model made 6-10 member integrations to minimize the weather noise and enhance the climate signal. A multi-model ensemble mean was made to reduce uncertainties arising from the models' parameterization of subgrid scale processes.

Wang et al (2005), however, demonstrated that climate modeling and prediction by prescribing the lower boundary conditions was inadequate when dealing with the particular phenomenon of the Asian summer monsoon rainfall. They showed that an AGCM which can simulate realistic SST-rainfall relationships when coupled with an ocean model, fails when forced by the same SSTs that are generated in its coupled run. Observed seasonal mean rainfall and SST anomalies are negatively correlated in heavy rainfall regions of the Asian summer monsoon. Furthermore, SST anomalies trail behind the rainfall by a month, suggesting that it is the SST anomalies which are forced by the atmospheric anomalies. This means that for a good monsoon prediction, the atmospheric feedback on the SST is critical.

In general it is clear that the current models are unable to simulate the mean monsoon climate or its intraseasonal variations and therefore they also cannot realistically simulate the interannual variability of the monsoon and and monsoon climate anomalies. The complex geography of the south Asian monsoon region results in regional rainfall peculiarities which can perhaps be better handled by very high resolution models. However, the large discrepancies between the observed and simulated rainfall point to basic deficiencies in the physical parameterization schemes and once again suggest the use of coupled models.

Sajani et al (2007) examined the fidelity of the AGCM of the Meteorological Research Institute (MRI), Japan, run in an ensemble mode with observed SST, in simulating Indian summer monsoon rainfall and its interannual variation. While the simple ensemble mean captured the essential features of the rainfall climatology and its extreme anomalies, it still showed a systematic bias in simulating the mean seasonal variation of rainfall over the Asia-Pacific region. When this bias was removed the model was able to capture the precipitation response to fluctuations in SST boundary forcing more realistically.

## 4.13 Atmosphere-Ocean Coupled Models

It has been accepted since long that atmospheric processes cannot be predicted beyond a time limit of two weeks if they are considered in isolation (Charney et al 1981, Shukla 1998). Any prediction on a longer time scale can be attempted only with the involvement of land and ocean processes which are far slower and evolve over several months or even years. Atmospheric general circulation models (AGCMs) are basically designed for the purposes of short and medium range numerical weather prediction only. If they are run for longer periods, they would not know of the changing ocean conditions, and would have to be provided with SST and sea ice data, either as observed or predicted separately.

Ocean general circulation models (OGCMs) are the ocean counterparts of AGCMs. They deal with the ocean circulation and interior ocean processes, but again they have no knowledge of the atmosphere and have to be supplied with the necessary data about the temperature and other atmospheric parameters about the ocean surface.

To overcome the individual deficiencies of an AGCM and an OGCM, they should, ideally speaking, be coupled together. Such atmosphere-ocean general circulation models (AOGCMs) become extremely complex and demand huge computer resources. As a tradeoff, they can be run only on a coarse resolution of say 500 km in the horizontal. However, many important local features, such as orography-induced rainfall in the monsoon region, are lost because of the coarse resolution. AOGCMs are the recommended tool for climate prediction, as we shall see in the next chapter.

As of now, there is no single AGCM that can produce realistic simulations or predictions of the monsoon in all aspects and most of the results are model-dependent. This takes us into a ridiculous situation wherein we may be required to first know how the monsoon was likely to behave and then choose a model that would suitable for that expected behaviour!

## 4.14 References


Ananthakrishnan R., Acharya U. R. and Ramakrishnan A. R., 1967, "On the criteria for declaring the onset of the southwest monsoon over Kerala", *Forecasting Manual, FMU Report No. IV-18.1*, India Meteorological Department, Pune, 52 pp.

Ananthakrishnan R., Srinivasan V. Ramakrishnan A. R. and Jambunathan R., 1968, "Synoptic features associated with onset of southwest monsoon over Kerala", *Forecasting Manual, FMU Report No. IV-18.2*, India Meteorological Department, Pune, 52 pp.

Ananthakrishnan R. and Soman M. K., 1988, "The onset of the south-west monsoon over Kerala 1901-1980", *J. Climatology*, 8, 283-296.

Basu S., Ramesh K. J. and Begum Z. N., 1999, "Medium range prediction of summer monsoon activities over India vis-à-vis their correspondence with the observational features", *Advances in Atmospheric Science,* 16, 133-146.

Chakraborty A., Krishnamurti T. N. and Gnanaseelan C., 2007, "Prediction of the diurnal cycle using a multimodel superensemble. Part II: Clouds", *Monthly Weather Review,* 135, 4097- 4116.

Charney J. G. and Shukla J., 1981, "Predictability of monsoons", in *Monsoon Dynamics* [Ed: Lighthill J. and Pearce R. P.], Cambridge Univ. Press, New York, 99-109.

Chowdhury A. and Gaikwad S. D., 1983, "On some characteristic features of rainfall associated with monsoon depression over India", *Mausam*, 34, 33-42.

Chowdhury A., Urankar P. S. and Upadhye C. U., 1985, "Cloud characteristics of monsoon depressions as viewed by meteorological satellites", *Mausam*, 36, 491-498.

Das S., 2002, "Real time mesoscale weather forecasting over Indian region using MM5 modelling system". *Research Report No. NMRF/RR/5/2002*, National Centre for Medium Range Weather Forecasting, Noida, India, 75 pp.

Das S., 2003, "Mesoscale and cloud-resolving scale simulation of a heavy precipitation episode and associated cloud system using MM5 model", in *Weather and Climate Modelling* [Ed: Singh S. V. et al], New Age International Publishers, New Delhi, 106-117.

Das S., 2005, "Mountain weather forecasting using MM5 modelling system", *Current Science*, 88, 899-905.

De U. S. and Mukhopadhyaya R. K., 2002, "Breaks in monsoon and related precursors", *Mausam*, 53, 309-318.

Deepa R., Seetaramayya P., Nagar S. G., and Gnanaseelan C., 2007, "On the plausible reasons for the formation of onset vortex in the presence of Arabian Sea mini warm pool", *Current Science*, 92, 794-800.

Dhar O.N. and Mandal B.N, 1981, "Greatest observed one-day point and areal rainfall of India", *Pure Applied Geophysics*, 119, 922-933

Dwivedi S., Mittal A. K. and Goswami B. N., 2006, "An empirical rule for extended range prediction of duration of Indian summer monsoon breaks", *Geophysical Research Letters*, 33, doi:10.1029/2996GL027035.

Fasullo J. and Webster P. J., 2003, "A hydrological definition of Indian monsoon onset and withdrawal", *J. Climate*, 16, 3200-3211.

Gadgil S. and Sajani S., 1998, "Monsoon precipitation in the AMIP runs", *Climate Dynamics*, 14, 659- 689.

Gadgil S. and Joseph P. V., 2003, "On breaks of the Indian monsoon", *Proc. Indian Academy Science (Earth Planetary Sciences)*, 112, 529-558.

Gates W. L., 1992, "AMIP: The Atmospheric Model Intercomparison Project", *Bulletin American Meteorological Society*, 73, 1962-1970.

Godbole R. V., Kelkar R. R. and Murakami T., 1970, "Radiative equilibrium temperature of the atmosphere along 80° E longitude", *Indian J. Meteorology Geophysics*, 21, 43-52.

Goswami B. N. and Xavier P. K., 2003, "Potential predictability and extended range prediction of Indian summer monsoon breaks", *Geophysical Research Letters*, 30, DOI: 10.1029/2003GL017810.

Goswami B. N. and Xavier P. K., 2005, "ENSO control on the South Asian monsoon through the length of the rainy season", *Geophysical Research Letters*, 32, DOI: 10.1029/2005GL023216.

Gupta G. R. and Onkari Prasad, 1991, "Activity of southern Indian Ocean convergence zone as seen in satellite cloud data during pre-monsoon months", *Mausam*, 42, 145-150.

IMD, 2005, *Climate Diagnostics Bulletin,* India Meteorological Department, Pune, July 2005.

Jenamani R. K., 2004, "Distinct synoptic patterns associated with pre-break onset phase and revival of normal monsoon phase", *Mausam*, 55, 591-598.

Joseph P. V., Eischeid J. K. and Pyle R. J., 1994, "Interannual variability of the onset of the Indian summer monsoon and its association with atmospheric features, El Nino, and sea surface temperature anomalies", *J. Climate*, 7, 81-105.

Joseph P. V. Sooraj K. P. and Rajan C. K., 2003, "Conditions leading to monsoon onset over Kerala and the associated Hadley cell", *Mausam*, 54, 155-164.

Joseph P. V., Sooraj K. P. and Rajan C. K., 2006, "The summer monsoon onset process over South Asia and an objective method for the date of monsoon onset over Kerala", *Int. J. Climatology*, 26, 1871-1893.

Joshi P. C. and Simon B., 1994, "NOAA/TOVS derived upper tropospheric temperature changes associated with the onset of southwest monsoon over Kerala coast", *Mausam*, 45, 155-160.

Kalsi S. R. and coauthors, 2004, "Various aspects of unusual behaviour of monsoon 2002", *Meteorological Monograph No. Synoptic Meteorology 2/2004*, India Meteorological Department , Pune, 105 pp.

Kang I. and coauthors, 2002, "Intercomparison of the climatological variations of Asian summer monsoon precipitation simulated by 10 GCMs", *Climate Dynamics,* 19, 383-395.

Kang I. and Shukla J., 2005, "Dynamical seasonal prediction and predictability of monsoon", in *The Asian Monsoon* [Ed. Wang B.], Springer Praxis Books, 580-612.

Kesarkar A., Ratnam J. V., Dalvi M., Kaginalkar A., 2005, "Real time weather forecasting experiments using WRF models at C-DAC, India", *6th WRF / 15th MM5 Users' Workshop,* NCAR, USA, 27-30 June 2005.

Khole M. and De U. S., 2003, "A study of north-east monsoon rainfall over India", *Mausam*, 54, 419-426.

Khole M. and De U. S., 2004, "Association between the monsoon onset over Kerala (MOK) and sea surface temperature (SST) over north Indian Ocean", *Mausam*, 55, 495-496.

Krishnamurti T. N., Ardanuy P., Ramanathan Y. and Pasch R., 1981, "On the onset vortex of the summer monsoon", *Monthly Weather Review*, 109, 344-363.

Krishnamurti, T. N. and coauthors, 1999, "Improved weather and seasonal climate forecasts from multimodel superensemble", *Science*, 285, 1548-1550.

Krishnamurti, T. N. and coauthors, 2000a, "Multimodel ensemble forecasts for weather and seasonal climate, *J. Climate*, 13, 4196-4216.

Krishnamurti, T. N., Kishtawal C. M., Shin D. W. and Williford C. E., 2000b, "Improving tropical precipitation forecasts from a multianalysis superensemble", *J. Climate*, 13, 4217-4227.

Krishnamurti T. N., Gnanaseelan C. and Chakraborty A., 2007a, "Prediction of the diurnal cycle using a multimodel superensemble. Part I: Precipitation", *Monthly Weather Review*, 135, 3613-3632.

Madden R. and Julian P., 1971, "Detection of a 40-50 day oscillation in the zonal wind in the tropical Pacific", *J. Atmospheric Science*, 28, 702-708.

Madden R. and Julian P., 1994, "Observations of the 40-50 day tropical oscillation: A review". *Monthly Weather Revire*, 112, 814-837.

Mahajan P. N., 2003, "Satellite data for diagnostics of monsoon disturbances", *Mausam*, 54, 165-172.

Mahajan P. N. and coauthors, 2004, "Proper depiction of monsoon depression through IRS-P4 MSMR", *Proc. Indian Acadamy Science (Earth Planetary Sciences)*, 113, 223-243.

Mishra D. K., Gupta G. R., Onkari Prasad and Nath T., 1991, "Some aspects of southwest monsoon as seen in satellite cloud imagery", *Mausam*, 42, 261-264.

MM5, 2008, "MM5 modeling system overview", MM5 web site at http://www.mmm.ucar.edu/mm5/mm5-home.html.

Mohanty U. C., Mandal M., Das A. K. and Dimri A. R., 2003, "Mesoscale modeling of convective systems over India: Status and scope", in *Weather and Climate Modelling* (Eds: Singh S. V. et al), New Age International Publishers, New Delhi, 63-76.

Murakami T., Godbole R. V. and Kelkar R. R., 1970, "Numerical simulation of the monsoon along 80° E", *Proc. Conf. Summer Monsoon of Southeast Asia*, Norfolk, Va, USA, [Ed: Ramage C. S.]. 39-51.

Pai D. S. and Rajeevan M., 2007, "Indian summer monsoon onset: variability and prediction", *NCC Research Report No. 4*, India Meteorological Department, Pune, 25 pp.

Pearce R. P. and Mohanty U. C., 1984, "Onsets of the Asian summer monsoon 1979-1982", *J. Atmospheric Science*, 41, 1620-1639.

Prasad K., Kalsi S. R. and Datta R. K., 1990, "Wind and cloud structure of monsoon depressions", *Mausam*, 41, 385-370.

Raj Y. E. A., 1996, "Inter- and intra-seasonal variation of thermodynamic parameters of the atmosphere over coastal Tamilnadu", *Mausam*, 47, 259-268.

Rajagopal E. N. and coauthors, 2007, "Implementation of T254L64 Global Forecast System at NCMRWF", *NCMRWF Technical Report No. NMRF/TR/1/2007*, 42 pp.

Rajeevan M., Pai D. S. and Das M. R., 2001, "Asymmetric thermodynamic structure of monsoon depression revealed in microwave satellite data", *Current Scence.*, 81, 448-450.

Rajeevan M., Bhate J., Kale J. D. and Lal B., 2006, "High resolution daily gridded rainfall data for the Indian region: Analysis of break and active monsoon spells", *Current Science*, 91, 296-306.

Rakhecha P. R. and coauthors, 1990, Homogeneous zones of heavy rainfall of 1-day duration over India, *Theoretical Applied Climatology*, 41, 213-219.

Rama Rao, Y. V., Prasad K. and Sant Prasad, 2001, "A case study of the impact of INSAT derived humidity profiles on precipitation forecast by limited area model", *Mausam,* 52, 647-654.

Rama Rao Y. V., Hatwar H. R., Ahmad K. S. and Sudhakar Y., 2005, "An experiment using the high resolution Eta and WRF models to forecast heavy precipitation over India", *Pure Applied Geophysics*, 164, 1593-1615.

Ramesh K. J., Basu S. and Begum Z. N., 1996, "Objective determination of onset, advancement and withdrawal of the summer monsoon using large-scale forecast fields of a global spectral model over India", *J. Meteorology Atmospheric Physics*, 61, 137-151.

Rao R. R. and Sivakumar R., 1999, "On the possible mechanisms of the evolution of a mini-warm pool during the pre-summer monsoon season and the genesis of onset vortex in the south-eastern Arabian Sea", *Quarterly J. Royal Meteorological Society*, 125, 789-809.

Roy Bhowmik S.K. and Prasad K., 2001, "Some characteristics of limited area model precipitation forecast of Indian monsoon and evaluation of associated flow features", *Meteorol. Atmospheric Physics*, 76, 223-236.

Roy Bhowmik S.K., 2003, "Monsoon rainfall prediction with a nested grid mesoscale limited area model over Indian Region", *Proc. Indian Academy Sciences*, 112, 499-520.

Sajani S., Nakazawa T., Kitoh A. and Rajendran K., 2007, "Ensemble simulation of Indian summer monsoon rainfall by an atmospheric general circulation model", *J. Meteorological Society Japan*, 85, 213-231.

Sarkar R. P. and Chowdhary A., 1988, "A diagnostic structure of monsoon depression", *Mausam*, 39, 9-18.

Shenoi S. S. C., Shankar D. and Shetye S. R., 1999, "The sea surface temperature high in the Lakshadweep Sea before the onset of the southwest monsoon", *J. Geophysical Research*, 104, 703-712.

Shukla J., 1998, "Predictability in the midst of chaos: A scientific basis for climate forecasting", *Science*, 282, 728-731.

Shyamala B., 2005, Personal communication.

Sikka D. R. and Gadgil S., 1980, "On the maximum cloud zone and the ITCZ over Indian longitudes during the southwest monsoon", *Monthly Weather Review*, 108, 1840-1853.

Simon B. and coauthors, 2001, "Monsoon onset-2000 monitored using multi-frequency microwave radiometer on-board Oceansat-1", *Current Science*, 81, 647-651.

Soman M. K. and Krishna Kumar K, 1993, "Space-time evolution of meteorological features associated with the onset of the Indian summer monsoon", *Monthly Weather Review*, 121, 1177-1194.

Srinivasan V., Raman S. and Ramakrishnan A. R., 1971, "Monsoon depression as seen in satellite pictures", *Indian J. Meteorology Geophysics*, 22, 337-346.

Suresh R. and Raj Y. E. A., 2001, "Some aspects of Indian northeast monsoon as derived from TOVS data", *Mausam*, 52, 727-732.

Suresh R., Sankaran P. V. and Rengarajan S., 2002, "Atmospheric boundary layer during northeast monsoon over Tamilnadu and neighbourhood - a study using TOVS data", *Mausam*, 53, 75-86.

Vaidya S. S., 2006, "The performance of two convective parameterization schemes in a mesoscale model over the Indian region", *Meteorology Atmospheric Physics*, 92, 175-190.

Vaidya S. S., 2007, "Simulation of weather systems over Indian region using mesoscale models", *Meteorology Atmospheric Physics*, 95, 15-26.

Vinayachandran P. N., Shankar D., Kurian J., Durand F., Shenoi S. S. C., 2007, "Arabian Sea mini warm pool and the monsoon onset vortex", *Current Science*, 93, 203-214.

Wang B., Lin Ho, Zhang Y. and Lu M. M., 2004, "Definition of South China Sea monsoon onset and commencement of the east Asia summer monsoon", *J. Climate*, 17, 699-710.

Wang B., Ding Q., Fu X., Kang I., Kyung Jin K., Shukla J. and Doblas-Reyes F., 2005, "Fundamental challenge in simulation and prediction of summer monsoon rainfall", *Geophysical Research Letters*, 32, doi:10.1029/2005GL022734.

Webster P. J. and coauthors, 1998, "Monsoons: processes, predictability and the prospects for prediction", *J. Geophysical Research*, 103, C7, 14451-14510.

Webster P. J. and Hoyos C., 2004, "Prediction of monsoon rainfall and river discharge on 15-30 Day Time Scales", *Bulletin American Meteorological Society*, 85, 1745-1765.

WRF, 2008, Weather Research and Forecasting Model" web site at http://www.wrf-model.org/index.php.

Xavier P. K. and Goswami B. N., 2007, "An analogue method for realtime forecasting of summer monsoon sub-seasonal variability". *Monthly Weather Review*, 135, 4149-4160.

Zhang C., 2005, "Madden-Julian Oscillation", *Reviews of Geophysics*, 43, 1-36.

# Chapter 5

# Projection of Monsoon Behaviour in the 21$^{st}$ Century

The accuracy of weather forecasts based upon synoptic analysis and satellite imageries is best on the short range i.e. up to 48 hours. Medium range forecasts have improved substantially in recent years because of the progress in numerical modelling. However, the accuracy falls rapidly as the lead time is increased to a week or 10 days, and such forecasts have usually to be updated before their validity is over. It is therefore reasonable to ask how it is possible to make predictions on the seasonal or climate scale.

The explanation is that a degree of skill does exist in predicting anomalies in the seasonal average of the weather, or climate anomalies. This skill is not related to our inability to predict the day-to-day sequence of weather events during the seasonal period. The level of accuracy of seasonal predictions is higher than that based upon climatology alone, and arises from the slowly varying boundary conditions at the surface.

According to Hastenrath (1986) there are five different approaches to climate prediction: (1) extrapolation of empirically or theoretically deduced periodicities, (2) assessment of statistical relationships between rainfall and various meteorological elements, (3) finding a relationship of the rainfall at the height of the rainy season with the rainfall of the previous season, (4) diagnostic studies of climate and circulation anomalies and (5) numerical modelling.

The statistical approach to long range forecasting of monsoon rainfall has been described at length in Chapter 3. The problems in the prediction of the mean seasonal state of the monsoon by different numerical models, and the lack of coherence among them, have been reviewed in Chapter 4. In this chapter we shall consider the current efforts towards making the projections of the behaviour of the monsoon across the period of the 21$^{st}$ century. It is important to be aware of the skill of such predictions in the particular context of climate change arising of the current global warming trend.

### 5.1 Decadal Variability of Monsoon Rainfall

The southwest monsoon seasonal rainfall for India as a whole exhibits considerable year-to-year variations (Section 3.1, Figure 3.1.1). Even if the averaging of the AISMR is carried out over a longer period of say a decade, the variations do not get completely eliminated, but a residual pattern emerges to show alternating positive and negative anomalies (Parthasarathy et al 1994). This is referred to the decadal, multi-decadal or epochal variation of the monsoon rainfall.

**Table 5.1.1 Number of years of deficient and excess AISMR during different decades (Source: Guhathakurta et al 2006)**

| Decade | Number of years of deficient rainfall in the decade (AISMR < 8% of normal) | Number of years of excess rainfall in the decade (AISMR > 8% of normal) |
|---|---|---|
| 1901-1910 | 3 | 0 |
| 1911-1920 | 4 | 3 |
| 1921-1930 | 1 | 0 |
| 1931-1940 | 1 | 1 |
| 1941-1950 | 1 | 1 |
| 1951-1960 | 1 | 3 |
| 1961-1970 | 2 | 1 |
| 1971-1980 | 3 | 1 |
| 1981-1990 | 2 | 2 |
| 1991-2000 | 0 | 1 |
| 2001-2003 | 1 | 0 |

The most recent reanalysis of the Indian rainfall data has been performed by Guhathakurta et al (2006), who have shown that there is a clear signal of decadal variability in the monsoon rainfall (Table 5.1.1). The deficient (excess) monsoon years indicate those years in which when the percentage departures from the mean rainfall were less (more) than the standard deviation which was 8% of the mean. If these are clubbed together over successive 30-year periods, then the years 1901-1930 and 1961-1990 emerge as dry periods with the intervening years 1931-1960 as a wet period, and the years 1991-2020 as a possibly wet period. Guhathakurta et al also did a 31-year running averaging of the AISMR series to isolate the low-frequency behaviour and the curve shows the 30-year epochal trend in the monsoon rainfall (Figure 5.1.1).

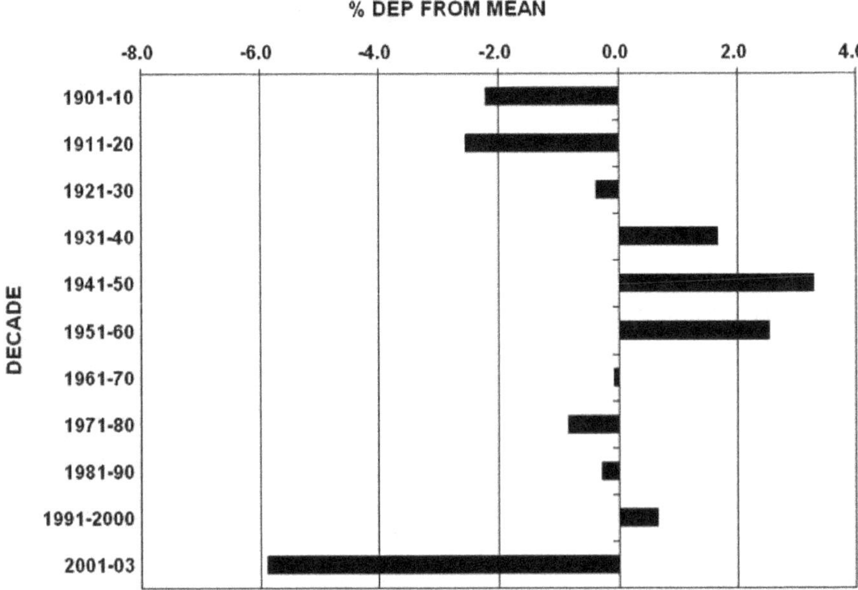

Figure 5.1.1 Decadal variability of AISMR in terms of the departure from normal (Source: Guhathakurta et al 2006).

Figure 5.1.2 31-year moving average of AISMR in terms of departure from normal (Source: Guhathakurta et al 2006)

## 5.2 Long Term Climate Trends over India

Whenever a continuous series of meteorological data gets generated over a long period of time, there is a natural scientific curiosity to find out whether the series is totally random or if it exhibits some kind of trend or periodicity. If the presence of a trend or oscillation is indeed discovered, it can be highly useful to extrapolate the time series, so that the future behaviour of the parameter may be foreseen.

One of the earliest discoveries of this type was with regard to the quasi-biennial oscillation (QBO) in the tropospheric zonal winds over the tropical and subtropical regions (Angell et al 1966) and in the surface temperature in many parts of the world (Landsberg et al 1963). Koteswaram and Alvi (1969) and Bhargava and Bansal (1969) had reported a QBO in the annual rainfall and monsoon season rainfall over some stations on the west coast of India.

Jagannathan and Parthasarathy (1973) made a very thorough statistical analysis of the trends and periodicities of annual rainfall over India at 48 stations having more than 70 years of continuous rainfall records up to 1967. As the various alternatives to randomness have the common property of low frequency variation, they calculated the serial correlation at small lags. The rainfall series was subjected to the Mann-Kendall rank statistic test to determine whether a significant trend existed or not, a Gaussian low pass filter was applied to eliminate the high frequency oscillations, and a power spectrum analysis was carried out to determine persistence. The results showed that there were increasing as well as decreasing tendencies in annual rainfall over many stations in India. These trends, however, were significant only at a few places which were not located in any particular region but distributed randomly. The QBO was also exhibited at several stations in the areas of increasing or decreasing trend and the 11-year solar cycle was also exhibited in both areas.

In the particular context of climate change and global warming, it is of great importance to investigate whether temperature and rainfall over India have been showing signs of an increasing or decreasing trend.


### 5.2.1 Precipitation

Guhathakurta et al (2006) have subjected the revised AISMR series for the 103-year period 1901-2003 to a low pass filter in order to suppress the high frequency oscillations. The weights used were 9-point Gaussian probability curves (0.01, 0.05, 0.12, 0.20, 0.24, 0.20, 0.12, 0.05 and 0.01). They found no evidence of any linear trend in this series. They also carried out a similar exercise for the rainfall of each of the monsoon months separately and applied linear regression technique and the Student's t-test. In an important conclusion, they saw no significant trend whatever in the all-India rainfall, either for the monsoon season as a whole, or for the four individual months in the monsoon season.

Guhathakurta et al (2006) also carried out a linear trend analysis to examine the long term trends in rainfall over different meteorological subdivisions and the monthly contribution of each of the monsoon months to annual rainfall. During the southwest monsoon season, 3 subdivisions viz., Jharkhand, Chattisgarh and Kerala, showed significant decreasing trend and 8 subdivisions viz., Gangetic West Bengal, West Uttar Pradesh, Jammu and Kashmir, Konkan and Goa, Madhya Maharashtra, Rayalaseema, Coastal Andhra Pradesh and North Interior Karnataka, showed significant increasing trends (Figure 5.2.1.1). The remaining 25 subdivisions did not exhibit any trend, either increasing or decreasing.

Kripalani et al (2003) who made an analysis of Indian monsoon rainfall data over a 131-year period (1871-2001) have come to the categorical conclusion that there is no clear evidence to suggest that the mean monsoon rainfall, or its decadal variability, or the frequency and intensity of extreme events, were in any way affected by global warming. There was no significant impact of global warming on monsoon rainfall that could be identified outside the natural climate variability.

Singh et al (2005) have viewed the rainfall trends from the perspective of 11 major river basins in India and 31 minor ones. They have found a decreasing trend from the 1960s in the rainfall over river basins in central India, while in other basins the rainfall has shown an increasing trend. Goswami et al (2006) have used the IMD gridded rainfall data set for 1951-2000 and shown that there is a significant increase in the frequency and intensity of extreme rain events in central India during this period. The trends in the precipitation extremes over India have been investigated by Joshi et al (2006) using rainfall data for 100 stations during 1901-2000. Most of the extreme rainfall indices have shown significant increasing trends at stations along the west coast of India and an increasing contribution of the extreme rainfall events to the monsoon seasonal rainfall.

A recent analysis of the rainfall gathered under the Global Precipitation Climatology Project (GPCP) from 1979 to 2005 has revealed the first signs of an increase in tropical rainfall (Gu et al 2007). The rainiest years in the tropics between 1979 and 2005 were mainly since 2001. The rainiest year was 2005, followed by 2004, 1998, 2003 and 2002, in that order. For the earth as a whole the total rainfall has changed very little, but in the tropics, where nearly two-thirds of all rainfall occurs, there has been an increase of 5%. However, the increase is concentrated over tropical oceans, with a slight decline over land. Over the 27 years between 1979 and 2005, some areas of the tropics experienced an increase in rainfall of as much as 0.5 mm/day per decade (red areas in Figure 5.2.1.2).

During El Nino years, total tropical rainfall did not change significantly but more rain fell over oceans than normal. The two major volcanic eruptions, El Chicon in Mexico and Mount Pinatubo in the Philippines, reduced overall tropical rainfall by about 5% during the following two years. With these effects removed from the rainfall record, the long term trend appears more clearly in the rainfall over both land and ocean.

### 5.2.2 Temperature

Figure 5.2.2.1 shows the all-India annual mean temperature anomalies for the period 1901-2007 with respect to the 1961-1990 mean. Over this period which is longer than a century, the annual mean temperature has exhibited an increasing trend of 0.5 °C/100 yr. However, if the spatial pattern of annual temperature anomalies (Figure 5.2.2.2) is examined, it is seen that there are some pockets of the western and eastern parts of the country where the trend has been negative. There are also some regions where there is no significant trend at all in the annual mean temperature.

It has been earlier thought that the warming over India was solely contributed by maximum temperatures. Kothawale et al (2005) have thrown some new light on the surface temperature trends over India. While the all-India mean annual temperature has shown a significant warming trend of 0.05 °C/10 yr during the period 1901-2003, the recent period 1971-2003 has seen a relatively accelerated warming of 0.22 °C/10 yr, which is largely due to unprecedented warming during the last decade. There is also a rise in the temperatures during the monsoon season, resulting in a weakened seasonal asymmetry of temperature trends reported earlier. The recent accelerated warming over India is observed both in maximum and minimum temperatures.

Sen Roy et al (2005) have analyzed the seasonal trends in the maximum and minimum temperature, diurnal temperature range and cloud cover over India for the time period 1931-2002 over a 1° × 1° lat./long. grid. They have also found that significant increases in maximum and minimum temperature have occurred over the Deccan plateau but the trends in the diurnal temperature range were not significant.

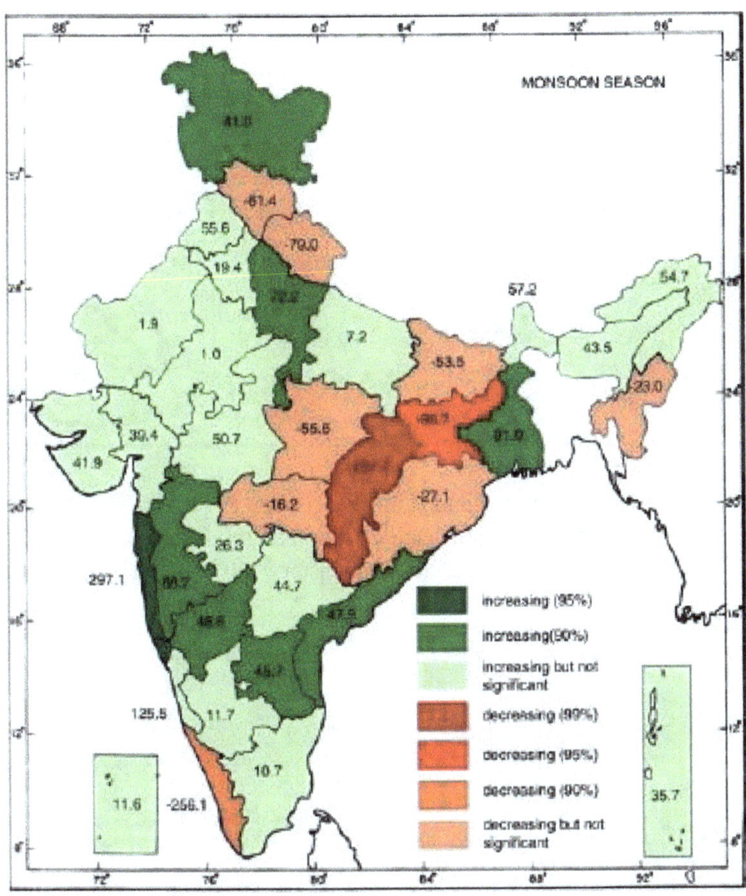


Figure 5.2.1.1 Increase/decrease in rainfall (mm/100 years) over the meteorological subdivisions of India for the southwest monsoon season. Different levels of significance are shaded (Source: Guhathakurta et al 2006)

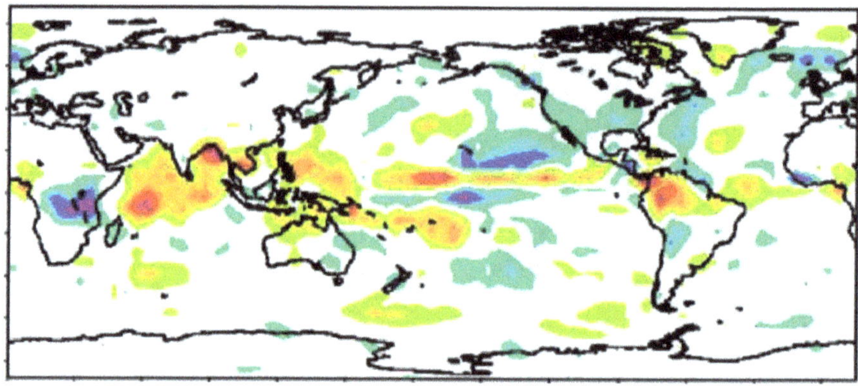

Figure 5.2.1.2 Change in tropical rainfall between 1979 and 2005. Increase (decrease) by 0.5 mm/day per decade shown as red (blue) area. (Source: http://www.nasa.gov/centers/goddard/news/topstory/2007/rainfall_increase.html)

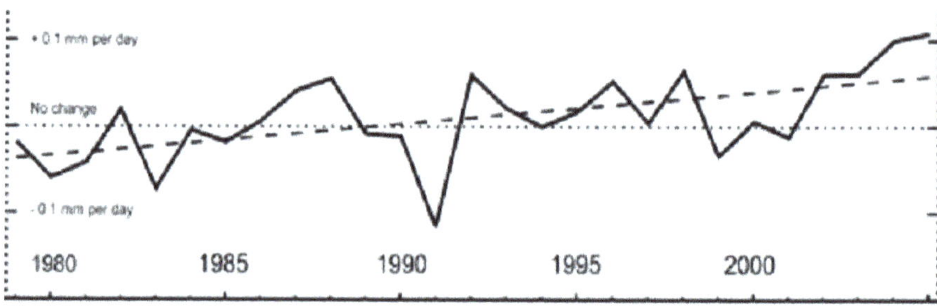


Figure 5.2.1.3 Change in tropical rainfall over the years 1979-2005. (Source: http://www.nasa.gov/centers/goddard/news/topstory/2007/rainfall_increase.html)

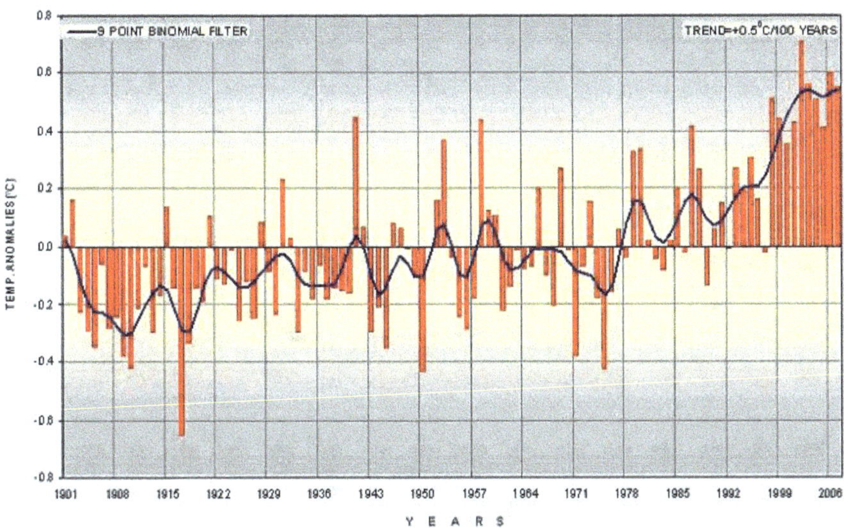


Figure 5.2.2.1 All-India annual mean temperature anomalies for the period 1901-2007 with respect to the 1961-1990 mean. The curve shows the subdecadal time scale variations smoothed with a 9-point binomial filter. There is an increasing trend of 0.5° C/100 yr. (Source: IMD)

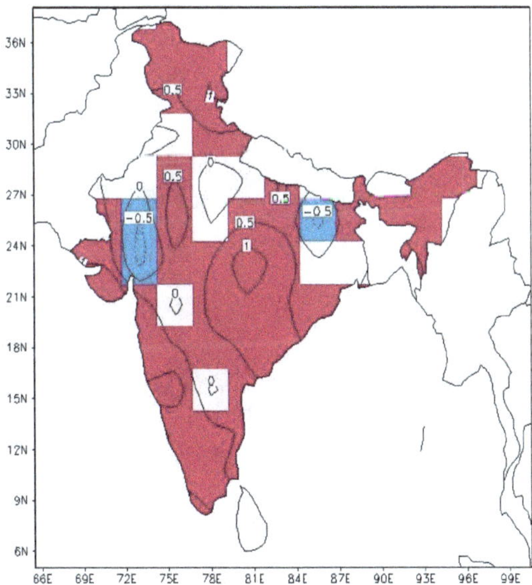


Figure 5.2.2.2 Annual mean temperature trends (°C/100 yr) over India for the period 1901-2007. Trends significant at 95% level are shaded red for positive trend and blue for negative trend. (Source: IMD)

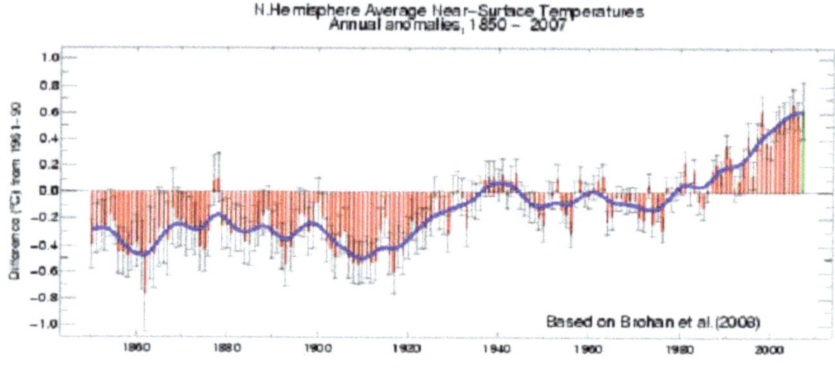

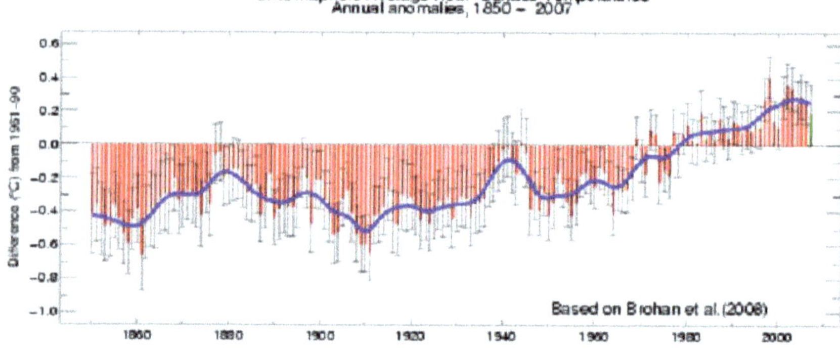

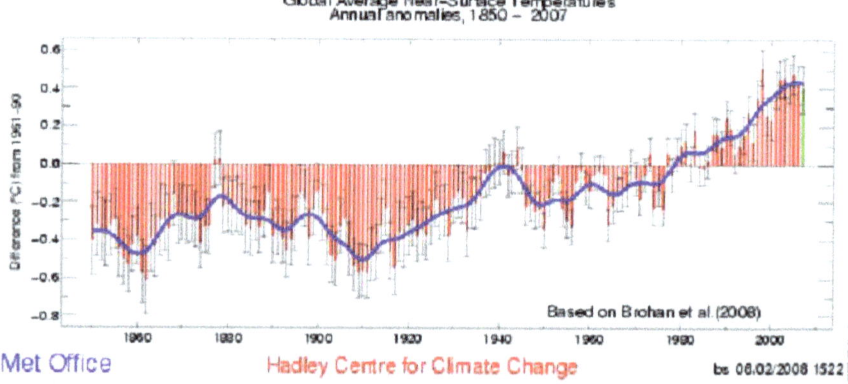

Figure 5.3.1 Annual northern hemispheric, southern hemispheric and global
anomalies of near-surface temperature during 1850-2007, with respect to the
1961-1990 mean along with their uncertainty estimates (Source:
http://www.metoffice.gov.uk/research/hadleycentre/obsdata/HadCRUGNS.html)

## 5.3 Global Warming

The World Meteorological Organization (WMO), has been issuing annual statements on the status of the global climate since 1993, that complement the assessments made periodically by the Intergovernmental Panel on Climate Change (IPCC).

Following established practice, WMO's global temperature analyses are based on two different data sets maintained by NOAA and the Hadley Centre using different methodologies. The Met Office Hadley Centre analysis showed that the global mean surface temperature in 2007 was 0.40 °C above the 1961-1990 annual average of 14 °C and hence marks the seventh warmest year on record. According to the NOAA/NCDC analysis, the global mean surface temperature anomaly was 0.55 °C above the 1901-2000 average of 13.9 °C which ranks 2007 as the fifth warmest year in its record. In fact, as the WMO is candid enough to acknowledge (WMO 2008), all the temperature values have uncertainties, arising mainly from gaps in the data coverage, which are such that the global average temperature for 2007 is statistically indistinguishable from each of the nine warmest years on record (Figure 5.3.1).

## 5.4 Emission Scenarios

A wide range of projections are presently available about the likely magnitude of global warming that would occur across the 21ˢᵗ century. The reason behind this divergence is that the various climate models currently in use vary greatly in their scope and complexity and the way in which they conceptualise future human activity.

The Special Report on Emissions Scenarios (SRES) was a report prepared by the IPCC for the Third Assessment Report (TAR) in 2001, on future emission scenarios to be used for running climate models. These have come to be known as the SRES scenarios.

Currently there are 40 such scenarios, each making different assumptions for future economic growth and technological development that would get reflected in greenhouse gas emissions, land use patterns and other anthropogenic forcings. These emission scenarios are grouped into families and future climate projections are often made in the context of a specific scenario family.

The A1 family of emission scenarios refers to a more integrated world characterized by rapid economic growth, a global population that reaches 9 billion in 2050 and then gradually declines, and the quick spread of new and efficient technologies. The A1FI scenario has an emphasis on fossil fuels, and A1T on non-fossil energy sources. The A1B scenario strikes a balance amongst all energy sources, not relying too heavily on one particular energy source, on the assumption that similar improvement rates apply to all energy supply and end use technologies.

The A2 family of scenarios represents a heterogeneous and more divided world consisting of independently operating, self-reliant nations, that preserve their local identities. There is a continuous growth in the population, regionally oriented economic development, slower and more fragmented technological changes and increase in the per capita income.

The B1 scenarios are of a more integrated, eco-friendly and stable world having the same population and economic growth as in A1, but with rapid changes towards a service and information economy, and introduction of clean and resource efficient technologies. The B2 scenarios visualize a world that is more divided, but more eco-friendly, population increasing at a slower rate than in A2 and intermediate levels of economic and technological development.

A1FI, A1B, A1T, A2, B1 and B2 are the six most commonly used emission scenarios in the context of future climate change.

**Table 5.5.1 Global Surface Temperature Change (°C)
during 2090-2099 relative to 1980-1999 (Source: Solomon et al 2007)**

| SRES Scenario | Best Estimate (°C) | Likely Range (°C) |
|:---:|:---:|:---:|
| B1 | 1.8 | 1.1-2.9 |
| A1T | 2.4 | 1.4-3.8 |
| B2 | 2.4 | 1.4-3.8 |
| A1B | 2.8 | 1.7-4.4 |
| A2 | 3.4 | 2.0-5.4 |
| A1FI | 4.0 | 2.4-6.4 |

## 5.5 Climate Projections for the 21ˢᵗ Century

Projections of global average surface warming for the end of the 21st century as computed by climate models are dependent, among other factors, upon the choice of the emission scenario. Of course, the warming that may actually occur will be determined by the emission scenario that would evolve in the real world over the next century. The projected global surface temperature rise during the period 2090-2099 compared to 1980-1999 for six SRES scenarios and for constant year 2000 concentrations are given in Table 5.5.1. The best estimates and corresponding likely ranges have been derived from a hierarchy of climate models, ranging from simple to complex AOGCMs.

While Table 5.5.1 gives the average values for the globe as a whole, Figure 5.5.1 shows the global distribution of the projected surface temperature changes for the early and late 21st century relative to the period 1980-1999. A significant feature of the figure is that the projected 21st-century temperature change is positive everywhere. It is greatest over land and at most high latitudes in the northern hemisphere during winter, and increases from the coasts towards the interior regions of the continents. In geographically similar areas, warming is typically larger in arid than in moist regions. In contrast, warming is least over the southern oceans and parts of the North Atlantic Ocean.

What is most evident from Figures 5.5.1 and 5.5.2 is that the entire Asian monsoon region is likely to experience a warming accompanied by an increase in precipitation rate in the June-July-August season. Over many parts of India, especially the western peninsula and northeast India, the monsoon precipitation rate is likely to register an increase of the order of 10-20% over the current rate by the end of the century.

## 5.6 Climate Models

In the Fourth Assessment Report of the IPCC (Solomon et al 2007, Christensen et al 2007), it has been stressed that Atmosphere Ocean General Circulation Models (AOGCMs) should be used as the primary tool for capturing the global climate system behaviour. AOGCMs are useful for investigating the processes maintaining the general circulation and its natural and forced variability, for assessing the role of various forcing factors, and for projecting the future climate under likely scenarios of external forcing. When climate models are run to make projections several decades ahead, the initial conditions cease to matter, but the long term changes in radiative forcings assume a greater importance. However, over smaller spatial and temporal scales, the variability linked to weather is higher and the signal of climate change is smaller. Although AOGCMS are improving continuously, since there are no historical analogues for the perturbations to radiative forcing that may be expected over the 21st century, confidence in the models can be built only from indirect methods. Since the results are model-dependent, the spread within an ensemble of AOGCMs is taken as a measure of the uncertainty in projected future climate changes. Some regional responses are consistent across AOGCM simulations, but not all.

One of the tests of a climate model is to see how faithfully it can simulate the present climate. It is equally important to test whether a model can simulate past climate variations. Model simulations of precipitation, sea level pressure and surface temperature have been generally improving, but deficiencies still remain particularly with respect to precipitation over the tropical regions and the simulation of clouds. Models generally tend to underestimate precipitation associated with extreme events.

Another way of testing a model is to use it for making seasonal runs with appropriate initial conditions so as to find out if it is adequately representing the key processes and teleconnections in the climate system.

Because of the coarse resolution of the AOGCMs, which is typically of the order of 500 km, many regional scale features tend to get missed out. To obtain useful results from AOGCMs below their grid scale, a process known as downscaling has to be resorted to. In dynamical downscaling, high-resolution regional models are run using boundary conditions that are either observed or obtained from AOGCMs. In statistical downscaling, empirical cross-scale relationships are developed from observed data and applied to the AOGCM results. Both approaches have their relative merits and demerits.

Under the aegis of the IPCC, a set of coordinated, standard experiments was performed by 14 AOGCM modelling groups from 10 countries using 23 models. The resulting multi-model simulations are called the MMD data set and most of them have been run with the A1B emission scenario. These were made available for analysis by hundreds of researchers worldwide. Multi-member ensembles from single models as well as multi-model ensembles have also provided useful results.

As per the Fourth Assessment Report of Working Group I of the IPCC (Christensen et al 2007), the annual mean temperature over south Asia is likely to rise by 3.3 °C by the end of the 21$^{st}$ century for the A1B scenario. The median warming varies seasonally from 2.7 °C in JJA to 3.6 °C in DJF, and is likely to increase northward in the area, and from sea to land in JJA (Figure 5.6.1), and more so during DJF.

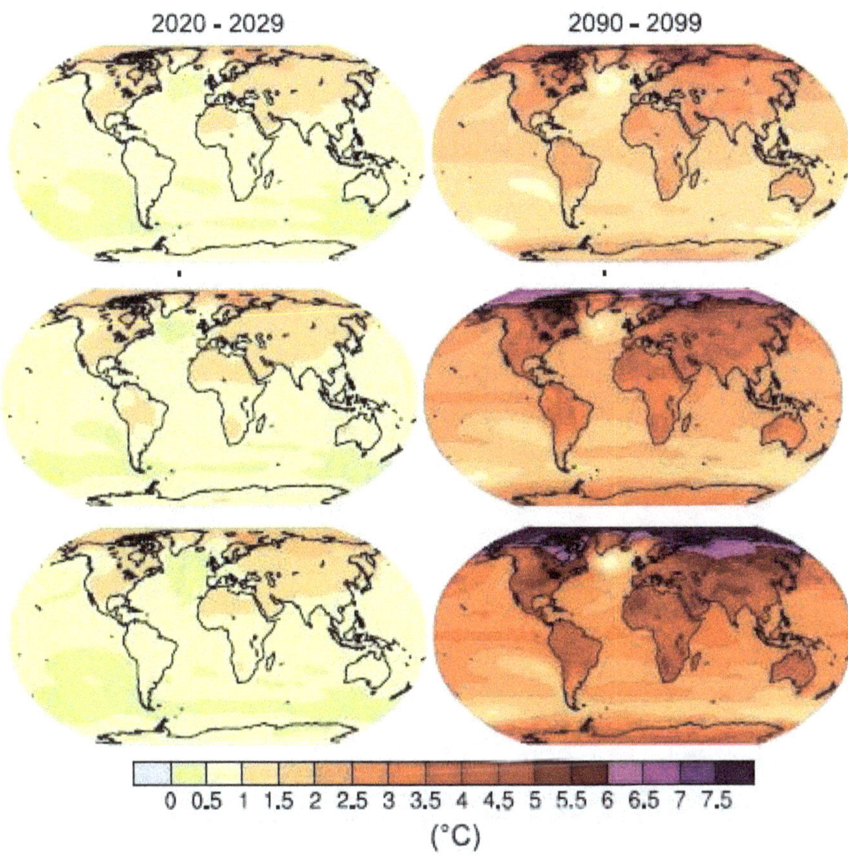



Figure 5.5.1 AOGCM multi-model average projections of the global distribution of surface temperature rise (°C) for the B1 (top), A1B (middle) and A2 (bottom) SRES scenarios averaged over the decades 2020 to 2029 (left) and 2090 to 2099 (right) relative to the period 1980-1999 (Source: Solomon et al 2007)

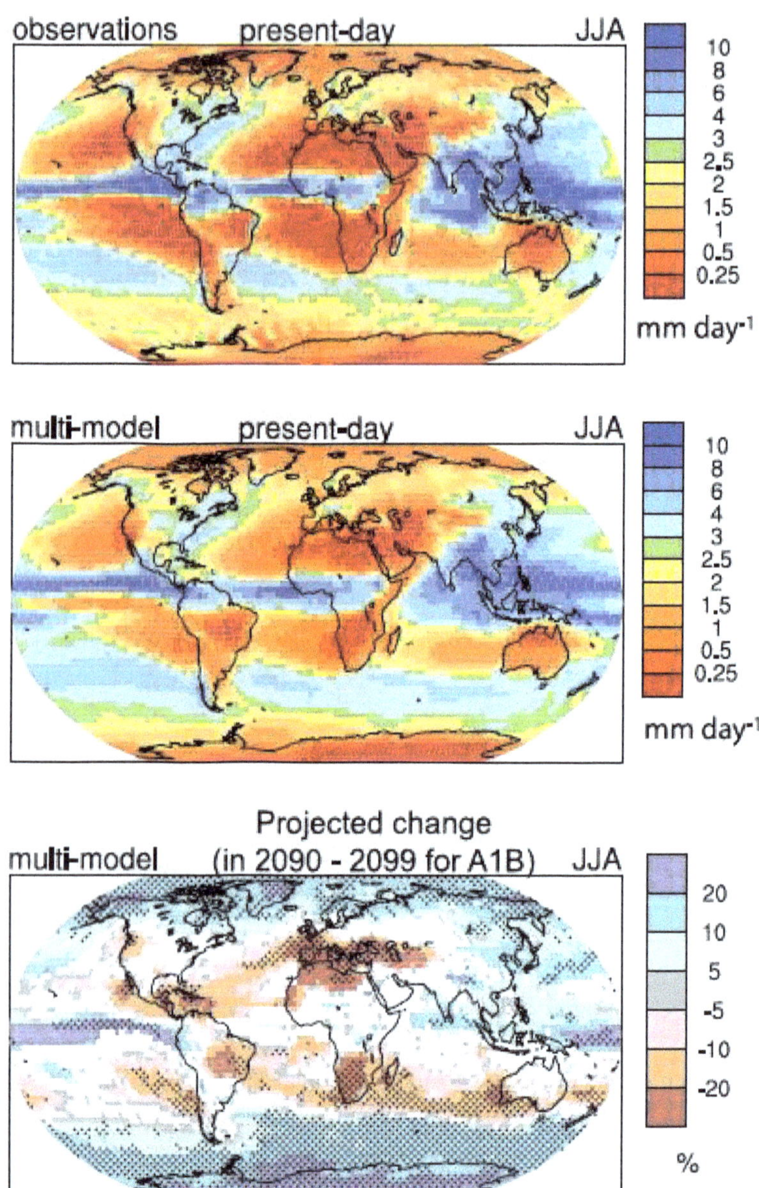



Figure 5.5.2 Global distribution of June-July-August mean precipitation rate in mm/day for the period 1979-1993: observed (top) and MMD A1B simulation (middle) and projected change in % (bottom) for the period 2090-2099 relative to 1980-1999. (Source: Solomon et al 2007)

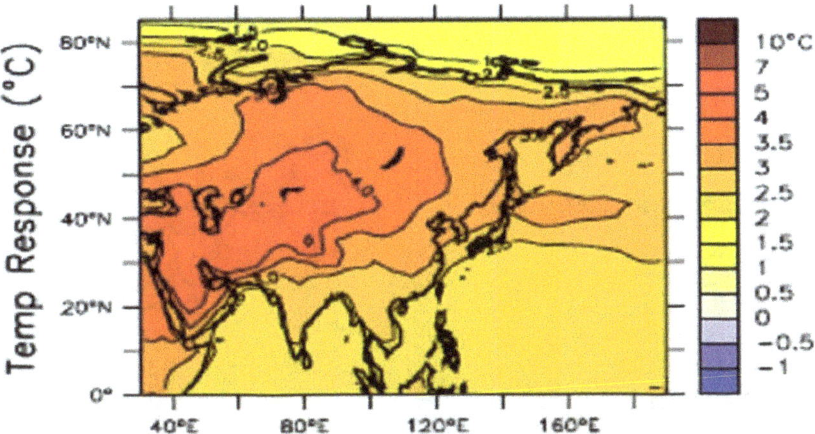

Figure 5.6.1 IPCC projections of temperature change (ºC) from 1980-1999 to 2080-2099 over Asia for June-July-August from the MMD A1B simulations averaged over 21 models. (Source: Christensen et al 2007)

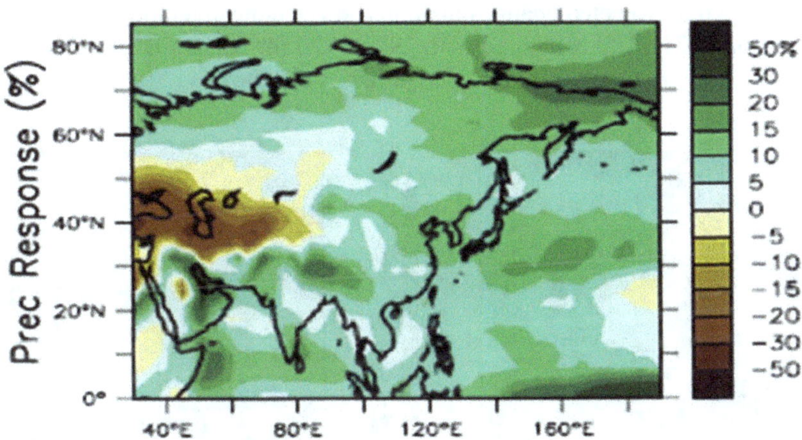

Figure 5.6.2  IPCC Projections of precipitation change (%) from 1980-1999 to 2080-2099 over Asia for June-July-August from the MMD A1B simulations averaged over 21 models. (Source: Christensen et al 2007)

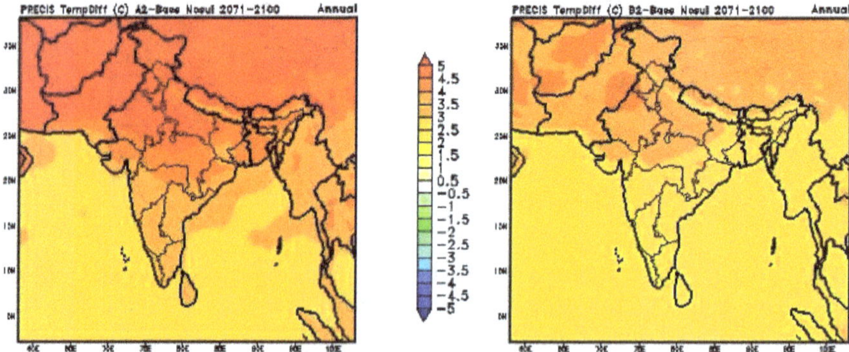

Figure 5.6.3 Projected changes in annual surface mean temperature in °C during 2071-2100 compared with 1961-1990 for A2 and B2 scenarios (Source: Rupa Kumar et al 2006)

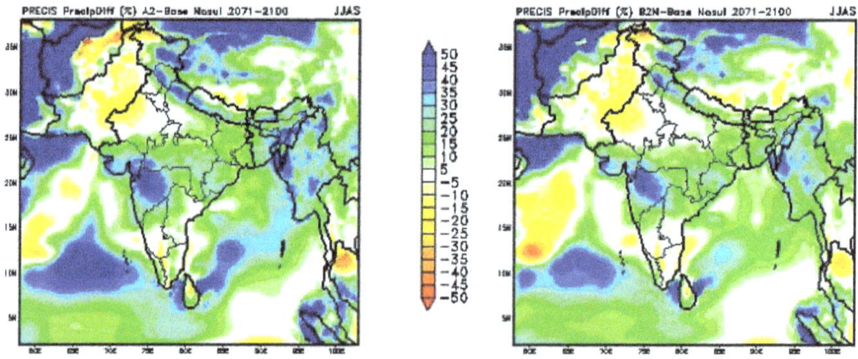

Figure 5.6.4 Projected changes in JJAS monsoon precipitation during 2071-2100 compared with 1961-1990 for A2 and B2 scenarios (Source: Rupa Kumar et al 2006)

The MMD models capture the general regional features of the monsoon, such as the low rainfall amounts coupled with high variability over northwest India. However, there has not yet been sufficient analysis of whether finer details of regional significance are simulated adequately in the MMD models. Most of the MMD-A1B models project a decrease in precipitation over south Asia in DJF (the dry winter season), and an increase during JJA (monsoon season) as seen in Figure 5.6.2. The median change is 11% by the end of the 21ˢᵗ century, and seasonally is -5% in DJF and 11% in JJA, with a large intermodel spread. Only 3 of the 21 models project a decrease in annual precipitation.

Recent work indicates that time slice experiments using an AGCM with prescribed SSTs, as opposed to a fully coupled system, are not able to accurately capture the South Asian monsoon response (Douville 2005). Thus, neglecting the short term SST feedback and variability seems to have a significant impact on the projected monsoon response to global warming, complicating the regional downscaling problem.

There are some studies (Kitoh and Uchiyama 2006) of the onset and withdrawal times of the Asian summer rainfall season in 15 MMD simulations. The results are indicative of a delayed withdrawal of the monsoon rains, while the onset dates are largely unaffected. This again amounts to an increase in the total monsoon rainfall. The results, however, are not clear and not very consistent among models.

As we have seen in Section 4.9.3, AGCMs are able to simulate the migration of the seasonal tropical rainbelt, but the monsoon rainfall is poorly simulated over the high rainfall regions along the west coast of India, over the north Bay of Bengal and adjoining northeast India. One reason is that the coarse resolution of some models is not able to capture the heavy rain that is associated with steep orography. However, the simulated annual cycles in South Asian mean precipitation and surface air temperature are reasonably close to those observed.

In a very recent work, Rupa Kumar et al (2006) have carried out regional climate model simulations for India, based on the second generation Hadley Centre regional climate model. This model is known as PRECIS (Providing Regional Climates for Impacts Studies) and it is a limited area high-resolution atmospheric and land surface model that is locatable over any part of the globe (Jones et al 2004). The boundary conditions are derived from a high-resolution AGCM (HadAM3H) with a horizontal resolution of 150 km × 150 km, in the so-called time slice experiments. Instead of running coupled models for century-long integrations, two time slices, namely 1961-1990 and

2071-2100, were selected from 240-year (1860-2100) long transient simulations with HadCM3. Ensembles of three baseline simulations for the period 1961-1990, three simulations for the A2 future scenario (2071-2100) and one simulation for the B2 future scenario (2071-2100) have been run with HadAM3H and assessed. PRECIS has been configured for a domain extending from about 1.5° N to 38° N and 56° E to 103° E with a resolution of about 50 × 50 km. The results are shown in Figures 5.6.3 and 5.6.4.

## 5.7 Likely Impacts of Global Warming on the Monsoon

PRECIS simulations under scenarios of increasing greenhouse gas concentrations and sulphate aerosols indicate a marked increase in surface air temperature in the 2071-2100 time frame as compared to the 1961-1990 baseline for both A2 and B2 scenarios, but the B2 scenario shows slightly less increase. The warming is expected throughout India, but there could be substantial spatial differences (Figure 5.6.3). Extremes in maximum and minimum temperatures are also expected to increase into the future, but with the night temperatures increasing more than the day temperatures.

In the PRECIS model simulations of Rupa Kumar et al (2006) described above, the mean monsoon rainfall for the baseline 1961-1990 was 939 mm with a standard deviation of 57 mm. Thus it has overestimated the mean rainfall while it has underestimated its variability. There were several other differences between the simulated and observed rainfall, particularly in quantitative terms. The mean simulated JJAS rainfall for 2071-2100 is 1114 mm for the A2 scenario and 1078 mm for the B2 scenario which translates into a 20% increase in the rainfall compared to 1961-1990.

The spatial distribution of the PRECIS simulations for the change in JJAS monsoon rainfall during 2071-2100 compared with 1961-1990 is shown in Figure 5.6.4. The maximum increase in rainfall of the order of 40% is seen over western Maharashtra and northeast India for both A2 and B2 scenarios. The increase is about 10-30% over most of central and eastern India and there is a decrease in rainfall of 5-10% over extreme northwest India. In other words, the projection is an accentuation of the present spatial rainfall variability over India with the wet regions getting wetter and the dry regions getting drier.

In a study by Tanaka et al (2005), intensities and trends of Hadley, Walker, and monsoon circulations were compared for the IPCC 20th Century simulations (20C3M) and for 21st century simulations (SRES A1B scenario), using the upper tropospheric (200 hPa) velocity potential data. In response to a global warming scenario, it is anticipated that the Hadley circulation may become weaker by 9%, Walker circulation by 8%, and monsoon circulation by 14% by the late 21$^{st}$ century as an ensemble mean of the IPCC model simulations. However, such results are to be viewed against the background of the poor capability of the current CGCMs in reproducing and predicting the tropical circulations. While a weakening of the monsoon circulation would suggest a reduction in monsoon rainfall, there are other model projections of an increased monsoon precipitation in a global warming scenario because of increased availability of total precipitable water in spite of a weaker circulation.

The response of the Asian summer monsoon to a possible doubling of $CO_2$ concentration has been the subject of many investigations in previous years However, these simulations were based upon single models chosen by the investigators and the results varied widely. Some studies showed that the monsoon was not likely to be affected at all, others projected a decline in monsoon precipitation, and many indicated that the monsoon rainfall could increase.

A recent concerted attempt in this domain of investigation has been that of Kripalani et al (2007), who have analysed the behaviour of 22 coupled climate models in the IPCC AR4 data base and their response to a doubling of $CO_2$. The study of Kripalani et al does not offer a complete evaluation of the performance of each of the models or their relative merits, but it is limited to the simulation of the total Indian southwest monsoon seasonal precipitation and its interannual variability (Tables 5.7.1 and 5.7.2). Out of the 22 chosen models, seven models simulated an annual cycle that was similar to the observed one in terms of shape and magnitude and six other models simulated the shape of the annual cycle well but underestimated the precipitation amounts, particularly during the spring and the summer periods. Another group of six models showed the rainfall maximum occurring a month later than observed, resulting in the underestimation of rainfall during spring and summer. The JJAS precipitation total varied between 50 and 91 cm across these 19 models while the coefficient of variation ranged from 3 to 13%. The remaining three models were unable to simulate the annual cycle to the desired degree of accuracy. Thus there is no single coupled climate model which can be regarded as ideal from the point of view of the Indian monsoon.

Six models were chosen for projection experiments with a 1% increase in $CO_2$ per year until the amount is doubled. This is about twice the rate of increase in $CO_2$ due to anthropogenic factors. The projected and simulated annual cycles for the monsoon derived from all six models show an increase in summer precipitation and perhaps an extension of the monsoon period. This may signify an intensification of the monsoon system. It is interesting to note that the simulated multi-model ensemble annual cycle accurately resembles that of the observed. The projected ensemble annual cycle clearly reveals an increase in precipitation during the summer and the following period but its simulation of observed precipitation is an underestimate. The projected increase in monsoon precipitation has a wide difference across the six models, ranging from 2.9% to 16.6%. The ensemble projects an 8% increase of JJAS precipitation from 81 to 87 cm.

## 5.8 A Rational View of Climate Change

The successive IPCC assessment reports have taken due note of the uncertainties in climate change science (Solomon et al 2007). Value uncertainties arise from data being inaccurate or not fully representative of the phenomenon of interest. Structural uncertainties arise from an incomplete understanding of the processes that control particular values or results, or inadequate model representation of the processes. The WMO has also pointed out the uncertainties in observations and their interpretation in its annual reports on the global climate (WMO 2008). It is not the intention here to go deeper into this aspect of uncertainty in global climate change studies and future projections, but only to discuss it in so far as it is relevant to the Indian monsoon.

The Indian southwest monsoon is one natural phenomenon that is important to a nation of 1 billion people. There is no doubt a need to make every attempt to foretell whether and how the monsoon is likely to change its behaviour over the $21^{st}$ century, particularly if the current trend of global warming continues relentlessly. Several possibilities have emerged out of diverse research investigations, which often get converted by the media into scary stories or doomsday forecasts. The common public cannot react to them in an objective manner while policymakers find it difficult to take definitive action in a largely uncertain scenario. It is therefore very necessary to a take a rational view about what we know with certainty and what we are uncertain about, at the present time.

We have seen in Chapter 2 that the Indian southwest monsoon has strong teleconnections with several global and regional land, ocean and atmospheric parameters and we have also seen in Chapter 3 how the teleconnections have a significant epochal variability. Therefore the monsoon would be affected by a climate scale change in such parameters, only if the teleconnections remain strong in the coming decades of this century. In the past, many teleconnections have been found and lost with time, but the monsoon has continued to be there and changed little. The AISMR graph from 1875 to 2007 does not exhibit any increasing or decreasing trend (Figure 3.1.1). A linear extrapolation of this graph would only imply that the long term average of AISMR is not likely to change drastically, at least in the immediate future. The pattern of the 30-year moving average of the AISMR which shows a small alternating rise and fall is currently showing a transition from a low rainfall epoch to a high rainfall epoch (Figure 5.1.2). This signal is another welcome indication of India having coming out of a drought phase and entering a phase of wet monsoons.

Figure 5.3.1 clearly indicates that the southern hemisphere has been warming at a slower rate compared to the northern hemisphere. This can be interpreted as an increased temperature contrast between the northern hemispheric landmass and the southern hemispheric ocean. If this contrast were to go on increasing, the Indian monsoon could be expected, with very simplistic reasoning, to thrive and intensify in the future.

Detailed inferences about the impact of global warming on the monsoon can, however, be drawn only with the climate models. Initial research work on the simulation of the circulation features and rainfall patterns of the Indian monsoon was mostly done in an isolated manner with individual atmospheric general circulation models of varying design and capabilities. These studies have more often than not, resulted in generating diverse and conflicting views of the future behaviour of the monsoon and its variability. It is only recently that a large number of atmosphere-ocean coupled general circulation models have been run in a coordinated way to assess their performance and suitability for monsoon prediction on a climate scale. In a qualitative sense, most of the climate models seem to agree on one point that the monsoon precipitation is likely to increase across the $21^{st}$ century in a global warming scenario with increasing $CO_2$. Thus the present consensus view emerging out of the climate model runs is that of an intensification of the monsoon, and there is a possibility that is not indicated by all models, that the monsoon season may extend somewhat longer than its current period.

However, the real difficulty arises when it comes to a quantification of the results. Most climate models are unable to simulate the observed features of the Indian monsoon in their totality. Many models underestimate the monsoon rainfall while some of them cannot simulate the observed monthwise precipitation pattern or the peak precipitation month. Hence only a few models can be trusted with the job of making a century scale prediction and even these have yielded diverse results (Table 5.7.1). There is no climate model currently available internationally that can be truly relied upon from all angles pertaining to the monsoon. The increase in precipitation resulting from a doubling of $CO_2$ in the atmosphere is likely to be anywhere from 2.9 to 16.6% for the A1B scenario (Table 5.7.2). This large uncertainty is further compounded by the possibility of the doubling of $CO_2$ not occurring in reality or the global $CO_2$ emissions not conforming to the A1B scenario.

In summary, all current indicators such as past statistical trends in AISMR, interhemispheric differences in the rate of global warming and projections of climate models, point to a strengthening of the Indian southwest monsoon along with the march of the 21st century, but there are large uncertainties when it comes to quantifying the likely change in monsoon rainfall.

## Table 5.7.1 Climate models participating in the IPCC AR4 experiments
### (Source: Kripalani et al 2007)

| No. | Originating group | Country | IPCC ID |
|---|---|---|---|
| 1 | Bjerknes Centre for Climate Research | Norway | BCCR-BCM2.0 |
| 2 | Beijing Climate Center | China | BCC-CM1 |
| 3 | National Centre for Atmospheric Research | USA | CCSM3 |
| 4 | Canadian Centre for Climate Modelling and Analysis | Canada | CGCM3.1 |
| 5 | Meteo-France/Centre National de Recherches Meteorologiques | France | CNRM-CM3 |
| 6 | CSIRO Atmospheric Research | Australia | CSIRO-Mk3.0 |
| 7 | Max Planck Institute for Meteorology | Germany | ECHAM5/MPI-OM |
| 8 | Meteorological Institute of University of Bonn/ METRI of KMA | Germany/ Korea | ECHO-G |
| 9 | LASG/Institute of Atmospheric Physics | China | FGOALS-g1.0 |
| 10 | NOAA/Geophysical Fluid Dynamics Laboratory | USA | GFDL-CM2.0 |
| 11 | NOAA/Geophysical Fluid Dynamics Laboratory | USA | GFDL-CM2.1 |
| 12 | NASAGoddard Institute for Space Studies | USA | GISS-AOM |
| 13 | NASA/Goddard Institute for Space Studies | USA | GISS-EH |
| 14 | NASA/Goddard Institute for Space Studies | USA | GISS-ER |
| 15 | Institute for Numerical Mathematics | Russia | INM-CM3.0 |
| 16 | Institut Pierre Simon Laplace | France | IPSL-CM4 |
| 17 | Center for Climate System Research (The University of Tokyo)/National Institute for Environmental Studies and Frontier Research Center for Global Change (JAMSTEC) | Japan | MIROC3.2 hires |
| 18 | Center for Climate System Research (The University of Tokyo)/National Institute for Environmental Studies and Frontier Research Center for Global Change (JAMSTEC) | Japan | MIROC3.2 medres |
| 19 | Meteorological Research Institute | Japan | MRI-CGCM2.3.2 |
| 20 | National Center for Atmospheric Research | USA | PCM |
| 21 | Hadley Centre for Climate Prediction and Research/ Meteorological Office | UK | UKMO-HadCM3 |
| 22 | Hadley Centre for Climate Prediction and Research/Meteorological Office | UK | UKMO-HadGEM1 |

**Table 5.7.2 Simulation of observed precipitation and projection for double CO$_2$ by various climate models (Source: Kripalani et al 2007)**

| IPCC ID of model | Simulation of observed precipitation features | | | JJAS precipitation amount (cm) | |
|---|---|---|---|---|---|
| | Shape of annual precipitation cycle | Monthly precipitation amounts | Month of peak precipitation | Simulation of observd amount | Projection for 2x CO$_2$ |
| BCCR-BCM2.0 | Well-simulated | Well-simulated | Well-simulated | 71 | |
| BCC-CM1 | Incorrect | Incorrect | Incorrect | - | |
| CCSM3 | Well-simulated | Well-simulated | Well-simulated | 78 | |
| CGCM3.1 | Well-simulated | Well-simulated | Well-simulated | 80 | 87 (+8.5%) |
| CNRM-CM3 | Well-simulated | Well-simulated | Well-simulated | 78 | 83 (+5.4%) |
| CSIRO-Mk3.0 | Not properly simulated | Underestimated in spring/summer | One month later than observed | 50 | |
| ECHAM5/MPI-OM | Well-simulated | Well simulated | Well-simulated | 74 | 77 (+2.9%) |
| ECHO-G | Well-simulated | Underestimated | Well-simulated | 60 | |
| FGOALS-g1.0 | Not properly simulated | Underestimated in spring/summer | One month later than observed | 62 | |
| GFDL-CM2.0 | Not properly simulated | Underestimated in spring/summer | One month later than observed | 61 | |
| GFDL-CM2.1 | Well-simulated | Well-simulated | Well-simulated | 76 | |
| GISS-AOM | Not properly simulated | Underestimated in spring/summer | One month later than observed | 57 | |
| GISS-EH | Well-simulated | Underestimated | Well-simulated | 63 | |
| GISS-ER | Well-simulated | Underestimated | Well-simulated | 61 | |
| INM-CM3.0 | Well-simulated | Underestimated | Well-simulated | 63 | |
| IPSL-CM4 | Incorrect | Incorrect | Incorrect | - | |
| MIROC3.2 hires | Not properly simulated | Underestimated in spring/summer | One month later than observed | 91 | 97 (+6.9%) |
| MIROC3.2 medres | Well-simulated | Well-simulated | Well-simulated | 90 | 97 (+8.2%) |
| MRI-CGCM 2.3.2 | Well-simulated | Underestimated | Well-simulated | 60 | |
| PCM | Incorrect | Incorrect | Two peaks | - | |
| UKMO-HadCM3 | Not properly simulated | Underestimated in spring/summer | One month later than observed | 72 | 84 (16.6%) |
| UKMO-HadGEM1 | Well-simulated | Underestimated | Well-simulated | 58 | |

## 5.9 References

Angell J. K., Korshover J. and Carpenter T. H., 1966, "Note concerning the period of the quasi-biennial Oscillation," *Monthly Weather Review*, 95, 319-323.

Bhargava B. N. and Bansal R. K., 1969, "A quasi-biennial oscillation in precipitation at some Indian stations,'' *Indian Journal* of *Meteorology and Geophysics,* 20, 127-128.

Christensen J. H. and coauthors, 2007, "Regional Climate Projections" in *Climate Change 2007: The Physical Science Basis. Contribution of Working Group I to the Fourth Assessment Report of the Intergovernmental Panel on Climate Change,* Cambridge University Press, Chapter 11, 847-940.

Douville H., 2005, "Limitations of time-slice experiments for predicting regional climate change over South Asia", *Climate. Dynamics.*, 24, 373-391.

Goswami B. N. and coauthors, 2006, "Increasing trend of extreme rain events over India in a warming environment", *Science*, 314, 1442-1445.

Gu G., Adler R. F., Huffman G. J. and Curtis S., 2007, "Tropical rainfall variability on interannual-to-interdecadal and longer time scales derived from the GPCP monthly product", *J. Climate*, 20, 4043-4046.

Guhathakurta P. and Rajeevan M., 2006, "Trends in the rainfall pattern over India", *Research Report No. 2/2006,* National Climate Centre, India Meteorological Department, Pune, 25 pp.

Hastenrath S., 1986, "On climate prediction in the tropics", *Bulletin American Meteorological Society,* 67, 696-702.

Jagannathan P. and Parthasarathy B., 1973, "Trends and periodicities of rainfall over India", *Monthly Weather Review*, 101, 371-375.

Jones R. and coauthors, 2004, *Generating high resolution climate change scenarios using PRECIS*, Hadley Centre for Climate Prediction and Research, Met Office Hadley Centre, UK, 40 pp.

Joshi U. R. and Rajeevan M., 2006, "Trends in precipitation extremes over India", *Research Report No. 3*, National Climate Centre, India Meteorological Department, Pune, 25 pp.

Kitoh A. and T. Uchiyama, 2006, "Changes in onset and withdrawal of the East Asian summer rainy season by multi-model global warming experiments", *J. Meteorological Society Japan*, 84, 247-258.

Koteswaram P. and Alvi S. M. A., 1969, "Secular trends and periodicities in rainfall at west coast stations in India," *Current Science,* 38, 229-231.

Kothawale D. R. and Rupa Kumar K., 2005, "On the recent changes in surface temperature trends over India", *Geophysical Research Letters*, 32, doi:10.1029/2005GL023528.

Kripalani R. H., Kulkarni A., Sabade S. S. and Khandekar M. L., 2003, "Indian monsoon variability in a global warming scenario", *Natural Hazards*, 29, 189-206.

Kripalani R. H., Oh J. H., Kulkarni A., Sabade S. S. and Chaudhari H. S., 2007, "South Asian summer monsoon precipitation variability: Coupled climate model simulations and projections under IPCC AR4", *Theoretical Applied Climatology*, 90, 133-159.

Landsberg H. E., Mitchell, J. M. Jr., Crutcher H. L. and Quinlan F. T., 1963, "Surface signs of biennial atmospheric pulse," *Monthly Weather Review*, 91, 549-556.

Parthasarathy B., Munot A., Kothawale D. R., 1994, "All-India monthly and seasonal rainfall series 1887–1993", *Theoretical Applied Climatology*, 49, 217-224.

Rupa Kumar K. and coauthors, 2006, "High-resolution climate change scenarios for India for the 21st century", *Current Science,*, 90, 334-345.

Sen Roy S. and Balling R. C., 2005, "Analysis of trends in maximum and minimum temperature, diurnal temperature range, and cloud cover over India", *Geophysical Research Letters*, 32, doi:10.1029/2004GL022201.

Singh N., Sontakke N. A., Singh H. N. and Pandey A. K, 2005, "Recent trend in spatiotemporal variation of rainfall over India – an investigation into basin-scale rainfall fluctuations" in *Regional Hydrological Impacts of Climate Change*, IAHS Publishers, 273-282.

Solomon S and coauthors, 2007, "Technical Summary" in *Climate Change 2007: The Physical Science Basis. Contribution of Working Group I to the Fourth Assessment Report of the Intergovernmental Panel on Climate Change*, Cambridge University Press, 74 pp.

Tanaka H. L., Ishizaki N. and Nohara D., 2005, "Intercomparison of the intensities and trends of Hadley, Walker and monsoon circulations in the global warming projections", SOLA, 2005, 77-80.

WMO, 2008, "WMO statement on the status of the global climate in 2007", *WMO No. 1031*, World Meteorological Organization, Geneva, 16 pp.

# Chapter 6

# Monsoon Prediction: Problems and Prospects

---

Since 1886, when IMD issued its first long range forecast of the southwest monsoon rainfall over India, until 2007, all of IMD's operational long range forecasts have been based only upon statistical models. Over this long period of more than a century, IMD has revised its models several times, adopted different statistical schemes, discarded many predictors and chosen new ones, changed the total number of parameters, covered different regions of the country, advanced the time of issue of the forecasts, and so on.

Compared to the situation that existed in 1886, when IMD was an eleven-year old organization, tremendous resources are now available in terms of access to global data, computational power, satellite imagery and satellite-derived products, and IMD has built up extensive data archives over the 133 years of its existence. Paradoxically, however, this has not been matched by a corresponding improvement in the prediction skill of long range forecast models. Rather, every successive model has sooner or later belied the promising expectations at the time of its adoption, and ended up bringing to the forefront, the inherent limitations of the statistical technique itself.

## 6.1 Limitations of Statistical Models

The multiple linear regression models for long range forecasting of monsoon rainfall take the general form of a single equation

$$AISMR = c_0 + c_1 p_1 + c_2 p_2 + ..... + c_N p_N$$

where $p_1$ to $p_N$ are the parameters that have been selected for the model, and $c_0$ to $c_N$ are the regression constants determined by the least squares technique so as to minimize the mean-squared differences between the actual and predicted AISMR for the period of the data series used. However, a major limitation of regression models, whether linear or nonlinear, is that the regression equation can be derived uniquely only in the case of a single-parameter model. If the number of predictors $N$ is more than one, it is possible to derive $2^N$-$1$ regression equations (Rajeevan et al 2006a, Delsole et al 2002). In the case of a 6-parameter model, for example, there can be $2^6-1$ or 63 possible regression equations. For a 16-parameter model, $2^{16}-1$ or as many as 65535 distinct multiple regression equations can be derived from the same pool of 16 parameters!

*Model Selection:* A fundamental problem encountered in the development of a statistical model of monsoon rainfall is that while a regression equation may be found to fit the past years perfectly, there is no guarantee that it would be able to predict the rainfall in future years equally well. It is therefore a practice to divide the data series into two segments: a training or development period and a testing or validation period. Various regression equations are built around the training or historical data set and tested with data during the testing period. From this exercise, it is possible to develop (a) the best model that explains the highest variance during the development period and (b) the best model that shows the highest predictive skill during the testing period. Ideally speaking, models (a) and (b) should be one and the same. In reality, however, a model has to be chosen so as to secure the maximum tradeoff between the two conflicting requirements of model design. The model that gets introduced for operational use is thus, at least to some extent, a compromise model. It is practically impossible to design a statistical model that can explain a 100% variance in the past AISMR, and at the same time, predict the future AISMR with a 100% confidence.

A historical review of IMD's long range forecasting models has shown that the model error has all along been of the order of ± 4 or 5 % or even higher, and in many individual years, the forecast errors have far exceeded the model error. In other words, there has not been any improvement over the years, in spite of the continuing attempts to revise the operational models based on increasingly rigorous and objective statistical methods (Gadgil et al 2004).

*Optimum Training Period:* Another consideration of prime importance in the development of long range forecast models, is the length of the data series that should be made use of. As far as rainfall is concerned, India is fortunate to have its earliest rainfall records dating back to the eighteenth century and all rainfall data for the country have been meticulously preserved at the IMD archives at Pune. A very reliable and continuous time series of the AISMR can therefore be constructed back to 1875 or even a little earlier to it, and statistical modellers can make use of it for regression analysis. However, a parallel data series is not available for every other parameter which they would like to correlate with the AISMR. The first pilot balloon stations were set up in the 1920s, and radiosonde stations in the 1940s and 1950s. Facsimile satellite APT imagery became available in the 1960s and INSAT imagery and INSAT-derived products only in the mid-1980s. Ocean data transmitted by Argo profiler floats are available just for the last few years.

The possible length of the data series that can be used for the regression analysis is thus, first of all, determined by the type of parameter that is chosen. This gets further reduced, as the recent years have to serve as the testing period of the model. It, however, does not follow that all the preceding years of data should be used for purposes of model development, and what constitutes an optimum period is a question that deserves serious attention.

Rajeevan et al (2006a) had adopted the criterion that the optimal length of the training period would be that for which the root mean square error (RMSE) of the forecasts for the period 1981-2004 was the lowest. For all the 63 possible models for the 6 predictors chosen by them, different training periods from 8 to 28 years were tried. The resulting scatter plot revealed that the RMSE of the models decreased with increase in the length of the training period, reached the minimum value around 23 years and then again increased with the length of the training period. Delsole et al (2002) used the strategy of selecting successive 25-year segments of the data series for predicting the rainfall in the year that followed immediately.

Even if recent studies suggest that 25 years is a good period to work with for long range forecasting models, questions can still be raised. For example, what would be better: to continue with a proven or trustworthy model for at least a few years, or to introduce every year a new and untested model that has been revised on the basis of the most recent 25 years' data? The choice is indeed difficult to make.

The other issue is related to the extreme monsoon years. If we examine the AISMR series from 1875 to date (Figure 3.1.1) it is easy to see that neither the drought years nor the excess rainfall years are uniformly spaced in time. There have been some 25-year periods during which drought occurrences were frequent and others with just one or two droughts. The regression model would obviously get tuned to the antecedent 25 years, and this would get reflected in the forecast for the 26th year. However, if there is an epochal change, the 26th year may not necessarily fit into the past 25-year pattern, and render the forecast incorrect. There is no easy solution to this problem.

*Optimum Number of Parameters:* How many parameters a long range forecast model of the monsoon rainfall should ideally have is another topic that has been debated extensively and perhaps inconclusively.

The earliest statistical long range forecast of the monsoon given by Blanford in 1886 was based upon just a single parameter, the Himalayan snow cover. The regression equation derived by Gilbert Walker on the basis of data from 1865 to 1903, and used by him for the IMD operational long range forecast for the year 1909, had four parameters: snowfall accumulation, Mauritius pressure and Zanzibar rainfall, all for the month of May, and the weighted average of pressure for the months March to May over Argentina and Chile in south America (Section 3.3).

However, Walker's subsequent work gives an impression that he was on an endless search for more and more predictors (Table 3.3.1). Gowariker et al (1989) while introducing a 15-parameter model had expressed the hope that with a larger set of parameters, the accuracy would increase. In fact they did add one more parameter to make it a 16-parameter model (Gowariker et al 1991). Later on, in hindsight, it was said that the number of parameters was not the issue, what really mattered was that the 16-parameter model had worked well from 1989 to 2001 (Gowariker 2002).

Delsole et al (2002) have argued that when the sample size is small compared to the number of parameters, the parameter values adapt themselves to the peculiarities of the sample. As a result, such models perform very well over the training data set but not necessarily on independent data of subsequent years. This is called the problem of artificial skill and it implies that regression models should restrict the predictors to the smallest number that is adequate for proper representation. In other words, it is a misconception that if the number of predictors is increased the prediction skill will improve, and doing so would only result in the past data having a better fit of the regression equation. In fact, if one goes on adding to the number of predictors, the model skill may improve at first, but at some point it will start yielding diminishing returns. Furthermore, in the atmosphere-land-ocean coupled system of the real world, most parameters would be correlated amongst themselves and it would impossible to generate a large set of parameters that are completely independent of one another, to fulfill the requirement of an ideal statistical model.

Delsole et al (2002) criticized the 16-parameter model of Gowariker et al (1991) as having artificial skill due to the large number of parameters. They showed with their own study that a model with good predictive skill could be built using only three predictors: the 500 hPa ridge location, Darwin pressure tendency and Eurasian surface temperature. Even a one-parameter model, using just the Eurasian surface temperature, would serve the purpose adequately.

After IMD revised its operational long range forecast models in the wake of the unprecedented drought of 2002, the number of parameters were brought down from 16 to 8-10 and further reduced to 3-6 in 2007 (Rajeevan et al 2004, 2006a). However, when the latest models were used operationally in 2007, their forecasts turned out to be in error by -12%.

Thus the issue of the optimum number of parameters to be used in long range forecasting models of the monsoon is yet far from resolved.

*Decline of Correlations with Time:* After the retirement of Gilbert Walker as the Director General of IMD in 1924, IMD continued to issue long range forecasts of the monsoon rainfall based on his models for two regions, northwest India and peninsular India. However, as the years went by, the performance of Walker's statistical model began to show a steady but definite deterioration (Montgomery 1940). The correlation coefficients of many of his parameters declined and even changed sign, and IMD had to make successive revisions to its operational long range forecasting schemes. For many years, the forecasts were framed in terms of qualitative expectations (Banerji 1950, Normand 1953, Jagannathan 1960). In 1989, IMD started the practice of giving quantitative forecasts for the country as a whole with the 16-parameter model (Gowariker et al 1989). Forecasts for homogeneous regions were reintroduced in 2000. The models were changed again in 2003 (Rajeevan et al 2004) and once again in 2007 (Rajeevan et al 2006a).

The periodical revisions of the operational models of IMD have of course been aimed at improving the models by bringing in more advanced statistical procedures. However, every revision has provided an opportunity to discard some parameters and introduce new ones. Figure 6.1.1 shows how the 21-year moving averages of the correlation coefficient of four popularly used parameters have changed with time and lost their predictive value. The 21-year correlation coefficient of the 500 hPa ridge position in April with the AISMR which had remained above 0.60 until 1977 (central year), fell to zero by 1987 (central year) and then changed sign from positive to negative. The correlation coefficient of the Darwin pressure tendency from January to April has shown an almost steady deterioration from -0.6 in 1980 to zero in 1993. Figure 6.1.1 is just illustrative of the temporal decline of the correlation coefficients, and the correlations of many other monsoon predictors show a change of sign from positive to negative and vice versa with time.

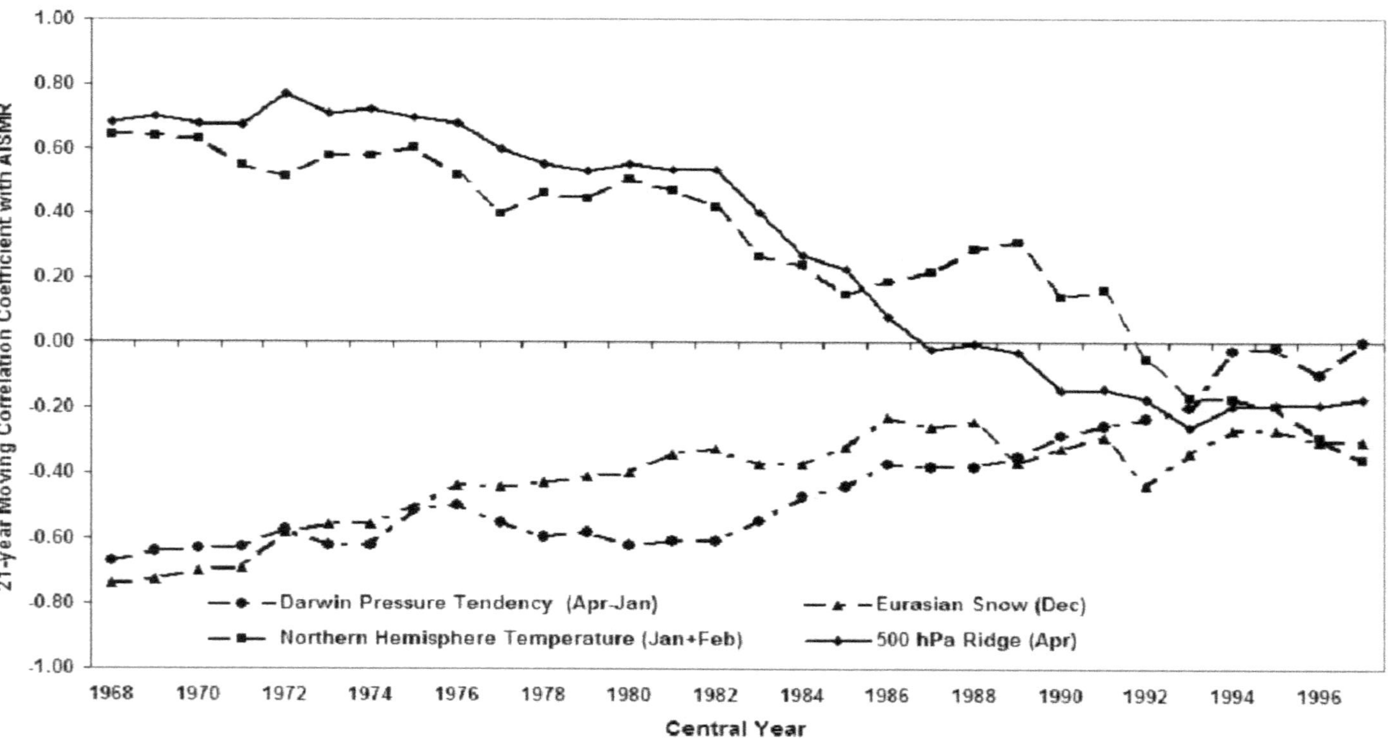


Figure 6.1.1 21-year moving correlation of different parameters with AISMR showing their decline and change of sign with time (Source: IMD)

It is not easy to explain all such epochal variations. Some of the changes could be purely statistical in nature, for not very clear reasons, with the physical linkages remaining in tact. In some cases, the physical relationships may themselves have undergone an epochal change which is reflected in the changing correlations. The performance of a model, which would obviously be the best at the time of its introduction into operational use, would thus get slowly eroded as the correlation coefficients decrease every following year. As a result, statistical long range forecasting models cannot hope to have a very long working life.

*Summary:* In spite of the problems and uncertainties associated with statistical models of long range forecasting of monsoon rainfall, which have been discussed in the above paragraphs, statistical models continue to be popular for several reasons. First of all they are far simpler than dynamical models and require very modest computational resources. Secondly, they do not demand a complete understanding of the physical processes of the monsoon, which in many respects, is yet far from complete. However, we are still far from the goal of designing an ideal statistical model and there is no unanimity yet on even the basic features like the number of parameters or the length of the training period that an ideal model should have. Further, the fundamental assumption that the statistical relationships which have been proved to exist in the past will also remain valid in the future, does not always hold good in reality. As a result, statistical models are, generally speaking, good at predicting normal monsoons but not the rainfall extremes which are much rare compared to instances of normal rainfall.

It is a common experience that whenever a statistical model is found to be successful, user expectations begin to rise and they grow beyond what the model can potentially deliver. Efforts have been made in the past to go beyond the AISMR and forecast the rainfall over smaller regions of the country or smaller periods within the monsoon season, like for instance, all-India rainfall for the month of July. However, as we deal with decreasing subseasonal or subregional scales, we have to work with increasingly smaller correlation coefficients, and a statistical prediction cannot be attempted beyond certain limits. The correlations of rainfall with global predictors fall rapidly towards zero as the spatial domain reduces to the size of a taluka or village, or as the time scale reduces to a week or pentad. No statistical model can ever be built to predict in the last week of May, the day-to-day pattern of rainfall during the entire monsoon season over an individual farmer's field. This limitation of statistical models is something that users have to understand and accept.

In short, statistical forecasting of monsoon rainfall still remains a challenging task. In the most recent paper on this subject (Xavier et al 2007a), a sense of 'futility' is very evident. They have argued that there is an intrinsic limit to predicting the all-India seasonal mean monsoon rainfall, but that even a skillful prediction may not be of much use in agricultural and water management applications. That there are serious limitations to the statistical models is very well-accepted; in fact these limitations have been discussed elaborately in this section. But it is difficult to concede to the viewpoint that forecasts of the AISMR are generally not useful. We have seen in sections 1.6 and 1.7 how the annual agricultural production of the country, and therefore the GDP, are closely linked to the AISMR. Long range forecasts of AISMR provide an early indication of the general behaviour of the southwest monsoon that can be expected, and they are vital for purposes of large scale advance planning. They are not intended to serve the farmer in his field, for whom other types of specialized agrometeorological services are being provided.

### 6.1.1 New Compilations of Indian Rainfall Data Series

The availability of an authentic, quality-controlled, well-maintained, standardized and continuous rainfall data series is at the heart of any successful attempt to predict monsoon rainfall, whether through statistical or dynamical models. While IMD is the custodian of all the massive rainfall data archives for the country, the most widely used Indian rainfall data set is that which was compiled by scientists of the Indian Institute of Tropical Meteorology, Pune from the basic IMD data and made open for use in scientific research (Mooley et al 1981, Parthasarathy et al 1987, Parthasarathy et al 1995, IITM 2008). This has come to be known popularly as the 'Parthasarathy data set' and it is based upon the rainfall data of 306 stations available from 1871. It, however, excludes the hilly regions of the country in view of their meagre rain-gauge network and the poor areal representation of a rain-gauge in a hilly terrain. Two island subdivisions far away from the mainland have also not been included. Thus, the contiguous area having a network of 306 stations over 30 meteorological subdivisions measures about 2,880,000 sq km, which is about 90% of the total area of the country. Since hill stations always receive copious rains compared to the plains, the AISMR derived from the Parthasarathy data set is a little less than what would be derived if all the meteorological subdivisions were taken into consideration.

In a very significant effort, IMD has now come out with a new homogeneous rainfall data set spanning a time period longer than a century, from 1901 to 2003, over an extensive and dense network of 1476 rain gauges across the country (Guhathakurta et al 2007). The analysis includes stations in the hilly regions, and excludes stations which had a break of more than 10% in the data series. The network is so chosen that there are at least two stations in each of the 458 meteorological districts of the country, so that districtwise rainfall statistics can be generated.

In yet another major initiative, IMD has also produced a high resolution ($1° \times 1°$ lat./long.) daily gridded rainfall data set (Rajeevan et al 2006b). This is based upon 1803 stations which had a minimum 90% of data availability during the analysis period of 1951-2003. The new IMD gridded rainfall analysis has a more accurate representation of rainfall over the Indian region, especially along the west coast and northeast India. It will find extensive application in validation of climate and numerical weather prediction models and also for studies on intraseasonal variability and predictability of monsoon rainfall. The data set is likely to be updated to cover the period from 1901 onwards, in which case more than 100 years of gridded daily rainfall data will become available to the research community.

IMD deserves to be complimented for bringing out these two new updated compilations of Indian rainfall data, which are sure to spawn many new investigations of the variability of the Indian monsoon rainfall on smaller spatial and temporal scales than was hitherto possible.

### 6.1.2 Need to redefine the AISMR

The AISMR is the area-weighted average of the monsoon rainfall over the whole of India over a four-month period from 1 June to 30 September. However, whether such an averaging of rainfall across a country of subcontinental dimensions and which has some of the wettest as well as the driest places on earth, yields a realistic value, is a question that has not been given the attention that is deserves.

The main reason why the AISMR has been popularly used in investigations of monsoon rainfall is that it serves as a convenient norm for comparison of the results, both with statistical and dynamical models. Apart from this, there is, no physical or meteorological merit or advantage in the current definition of the AISMR.

First of all, although the average date of the onset of the southwest monsoon over Kerala is 1 June, the monsoon has arrived much earlier or later than this date in several years. Thus by adopting 1 June as the date of reckoning for the AISMR, we are either not taking into account the monsoon rainfall in the years of its early arrival, or including premonsoon thunderstorm rains in the years of late onset.

Secondly, it takes almost a month and a half for the southwest monsoon to cover the remotest parts of northwest India. However, as per the prevailing definition, all premonsoon rainfall over the country, which is characteristically different from the monsoon rainfall, and is associated with local thunderstorms and the passage of western disturbances, goes into the accumulation of the AISMR.

Thirdly, although the computation of the AISMR ends on 30 September, the monsoon rarely leaves on this precise calendar date. In many years, deficient monsoon rainfall gets compensated by a lingering monsoon and reservoirs may get filled up by good rains in October.

Thus, while the AISMR has served as a piece of useful and convenient statistical information, there is a strong case for the AISMR to be computed dynamically, considering the actual dates of onset and withdrawal of the monsoon over different parts of the country in different years. Such a dynamically computed AISMR would be more representative of the rainfall that is actually associated with the southwest monsoon and not vitiated by the inclusion of non-monsoon rainfall events.

It is heartening to note that scientists have now begun questioning the current definition of the AISMR. A strong case for redefining the length of the monsoon season and allowing it to change interannually, has been made out by Xavier et al (2007b). They have shown that by arbitrarily limiting the monsoon season to 1 June to 30 September, the teleconnections of the monsoon with other climatic parameters could result in the misrepresentation of the physical processes. According to their analysis, it is conceivable that part of the apparent decrease in correlation between AISMR and ENSO indices, such as the Nino 3 SST is simply due to an inappropriate definition of the rainy season. It is hoped that the work of Xavier et al will trigger new investigations on this issue, lead to a better definition of the AISMR, and may even bring about an improvement in the statistical models through better teleconnections.

## 6.2 Changing Global Relationships of the Monsoon

The global and regional relationships of the Indian southwest monsoon, or the so-called teleconnections, have been discussed in depth in Chapter 2. These form the basis of statistical long range forecasting models as described in Chapter 3. A particular question that arises is whether in the scenario of climate change, these monsoon relationships or teleconnections as we know of today, are undergoing or likely to undergo a change themselves. If such signals are evident, then there would be a need to take a closer look at the changing teleconnections and accordingly design new long range forecast models for the monsoon.

## 6.2.1 Modes of Climate Variability

In the context of global warming it is important to note that surface temperatures over northern hemispheric land regions have warmed at a faster rate than over the oceans in the southern hemisphere. Long term changes have also been noticed in the atmospheric circulation, such as a strengthening of the midlatitude westerly winds and a poleward shift of the jet streams.

A significant component of regional climatic variability is known to be governed by fluctuations in the amplitude and sign of indices of a relatively small number of preferred circulation patterns, which are referred to as oscillations and modes. The most commonly known ocean-atmosphere coupled phenomenon of this type is the El Nino-Southern Oscillation (ENSO) over the equatorial Pacific Ocean, with a quasi-periodicity of 2 to 7 years (Sections 2.2 and 2.3). ENSO has been known to have global teleconnections, particularly with the Indian southwest and northeast monsoons. The lesser known North Atlantic Oscillation (NAO) is a measure of the pressure difference between the Icelandic Low and the Azores High, and of the westerly winds that connect them, especially in winter. Many observed variations in temperature, precipitation and the tracks of storms moving from the North Atlantic into Eurasia have been found to be associated with the NAO. In recent times, the relationship of the Indian monsoon with ENSO has been showing signs of weakening, whereas the distant NAO appears to be developing stronger bonds with the monsoon.

Several other modes and oscillations are known to be prevalent in the global atmospheric and ocean circulation (Solomon et al 2007a). The characteristics of fluctuations in the zonally averaged westerlies in the northern and southern hemispheres are referred to by their respective annular modes. The Northern Annular Mode (NAM) is a wintertime fluctuation in the amplitude of a pattern characterized by low surface pressure in the Arctic and strong midlatitude westerlies. NAM has links with the northern polar vortex and hence with the stratosphere. The observed strengthening of the midlatitude westerlies across the North Atlantic can be interpreted as being associated with the NAO or NAM and there is also an evidence of a multi-decadal variability in this regard.

The Southern Annular Mode (SAM) is the southern counterpart of NAM. It is the fluctuation of a pattern with low Antarctic surface pressure and strong midlatitude westerlies, but is present throughout the year. SAM is associated with long term changes in Antarctic temperatures. The Pacific-North American (PNA) pattern refers to an atmospheric large scale wave pattern featuring a sequence of tropospheric high- and low-pressure anomalies stretching from the subtropical west Pacific to the east coast of North America. The Pacific Decadal Oscillation (PDO) is a measure of the SSTs in the North Pacific, which in turn has a very strong correlation with the North Pacific Index (NPI). NPI is an index of the depth of the Aleutian Low.

## 6.2.2 ENSO

The year 1997 was characterized by the rapid development of an unprecedentedly intense El Nino. Going by the known influences of the El Nino on the southwest monsoon, it was expected that the monsoon would be very adversely affected. However, this did not happen and on the contrary, the AISMR for that year was 102% of the normal. Slingo et al (2000) investigated the causes of this anomalous response or rather the lack of repsonse of the monsoon to what was termed as the El Nino event of the century. They attributed this to the role played by the maritime continent in modulating the local Hadley circulation in addition to the influence of the El Nino on the Walker circulation. As a result in 1997, there was large scale convergence over the monsoon trough region and deep monsoon depressions over north India leading to above average precipitation.

However, further investigations of the peculiar behaviour of the monsoon against the strong El Nino of 1997, drew attention to the possibility that the long term relationship of the Indian monsoon with ENSO could be undergoing a change (Kripalani et al 1997, Krishna Kumar et al 1999). In a subsequent study, Kripalani et al (2003) made an extensive analysis of the data on Indian monsoon rainfall and other parameters associated with it over a 131-year period (1871-2001). They showed that the 11-year sliding correlation coefficient between the Indian monsoon rainfall and the Darwin pressure tendency had considerable secular variation. The relationship was weak up to 1930, strongly inverse during 1930-1970, and weak again thereafter. The correlation coefficient reduced to zero in the 1990s, and it changed sign from negative to positive after 1995.

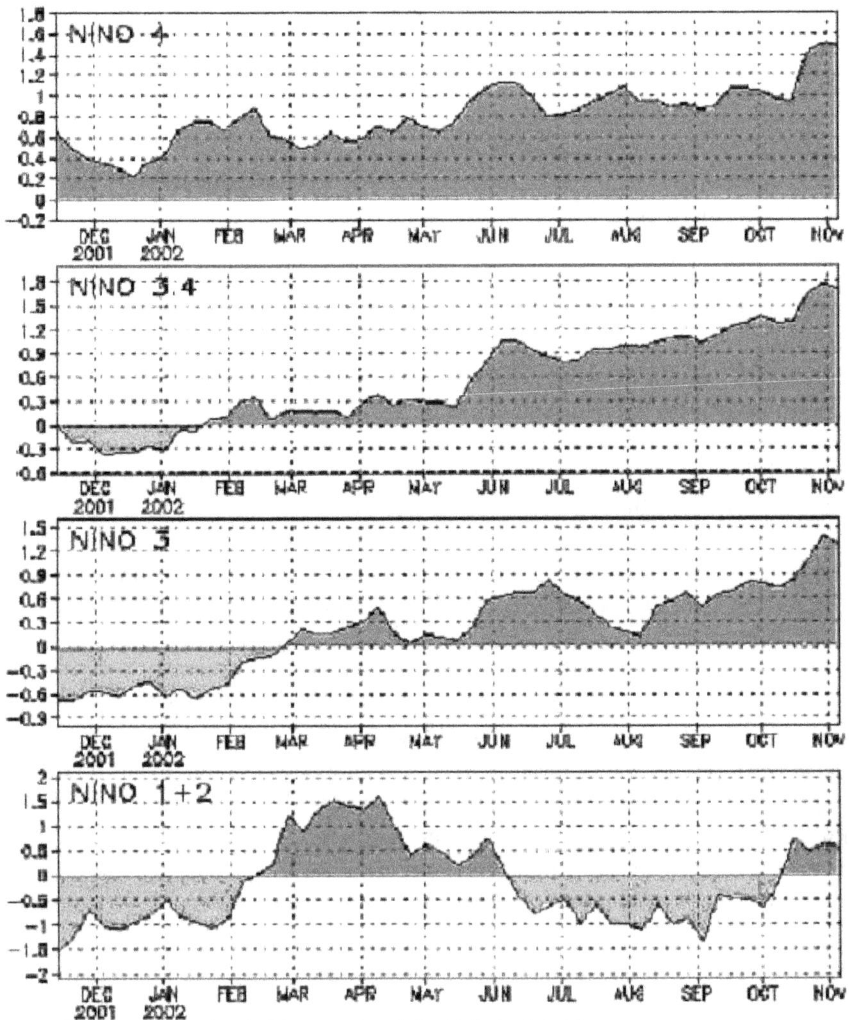

Figure 6.2.2.1 Evolution of SST anomalies over the four Nino regions during
the period December 2001 to November 2002
(Source: Climate Prediction Center/NCEP/NOAA web site)

In addition, Kripalani et al (2003) found that the relationship of the Indian
monsoon rainfall with northern hemisphere surface temperature which was
the strongest during 1960-1980 had also been weakening since 1980, and the
11-year sliding correlation had changed sign from positive to negative

around 1990 (see also Figure 6.1.1). Kripalani et al argued that if global warming was really the cause of the recent weakening of ENSO-monsoon relationship, then the relationship between monsoon rainfall and the northern hemisphere surface temperature should have strengthened and not weakened as observed. They concluded that the Indian monsoon variability appears to have delinked itself not only from the Pacific Ocean temperatures but also from the Northern hemisphere surface temperatures.

As a corollary to the weakening of the inverse ENSO relationship with the Indian southwest monsoon, there is a strengthening of the positive ENSO relationship with the Indian northeast monsoon (Pankaj Kumar et al 2007). This secular variation of the relationship may be due to the epochal changes in the tropospheric circulation associated with ENSO over the region.

The evolution of SST over the four Nino regions is not always in phase. In 2002, in particular, the Nino 1+2 SSTs were falling from May 2002 onwards and this was considered as a favourable predictor in the statistical model (Figure 6.2.2.1). However, SSTs over the other three Nino regions went on building throughout concurrently with the monsoon season of 2002 which was characterized by a major drought.

Krishna Kumar et al (2006) have hypothesized that the response of the monsoon rainfall to ENSO lies in the details of the tropical east Pacific Ocean warming. India is more prone to drought when the El Nino ocean warming signal extends westwards. The spatial configuration of the SST anomalies, or what have described as different El Nino flavours, has a significant impact on the Indian monsoon. Statistical models that use simple El Nino indices as predictors are liable to miss these flavours and go wrong.

## 6.2.3 ENSO Forecasting

Regardless of the indications that the relationship between the Indian southwest monsoon and ENSO may be slowly fading out, monsoon prediction cannot be done in isolation, overlooking what may be happening to ENSO. In fact, monsoon forecasters have a lot to gain from the recent global efforts towards a better understanding of the ENSO phenomenon and its prediction.

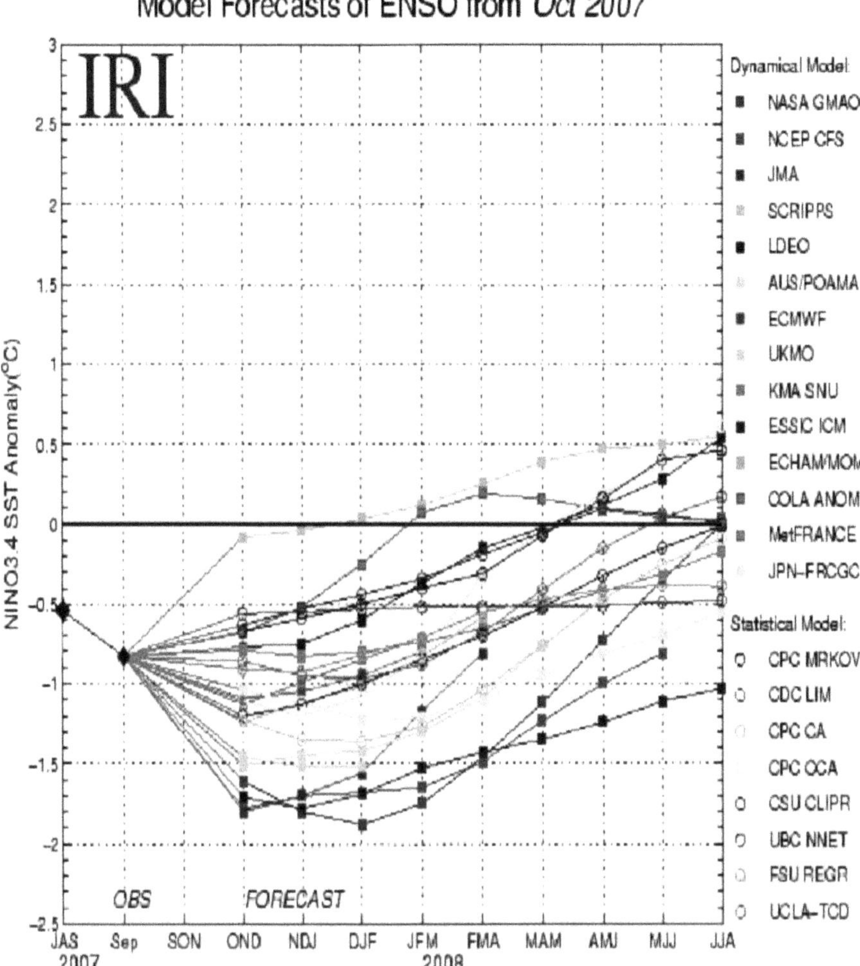


Figure 6.2.3.1 Forecasts of Nino 3.4 SST anomaly made in October 2007
with 22 different models for the next 10 months
(Source: http://iri.columbia.edu/climate/ENSO/currentinfo/SST_table.html)

For the purpose of ENSO forecasting, as many as 14 dynamical models are presently being run by agencies like NASA, NCEP, UKMO, ECMWF, ECHAM, COLA, MeteoFrance and JMA, while there are 8 statistical models of NOAA, NCEP, CPC, FSU and others. Forecasts of these models in terms of SST anomalies for the Nino 3.4 region for nine future overlapping 3-month periods are posted every month on the internet by the U.S.-based International Research Institute for Climate and Society, formerly known as the International Research Institute for Climate Prediction.

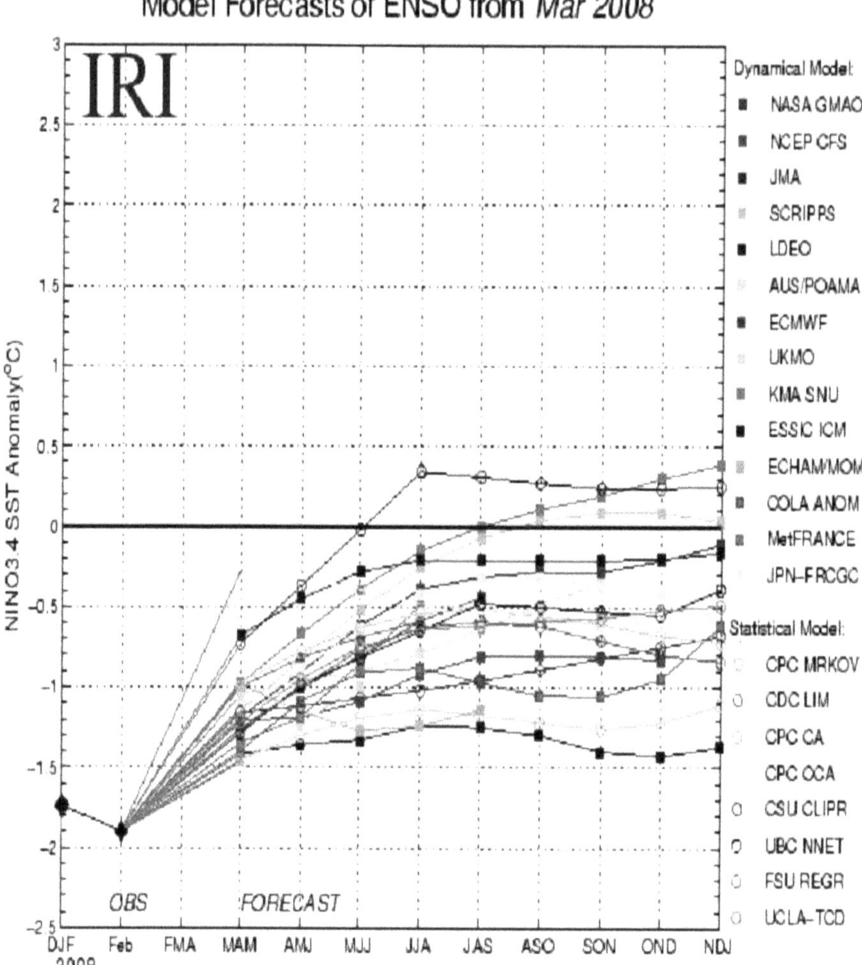



Figure 6.2.3.2 Forecasts of Nino 3.4 SST anomaly made in March 2008
with 22 different models for the next 10 months
(Source: http://iri.columbia.edu/climate/ENSO/currentinfo/SST_table.html)

Figure 6.2.3.1 shows the family of curves arising out of such forecasts made
in October 2007 for the next 10 months. Figure 6.2.3.2 is a similar diagram
showing the forecast plume for March 2008. What is at once evident is the
scatter among the curves, which is so large that it is called a forecast plume.
The scatter among the various forecasts is of the order of 2 °C and persists
throughout the forecast period. The forecasts indicate a range of possible
scenarios ranging from neutral to strong La Nina conditions. Moreover, the

October 2007 forecasts of SST anomalies for February 2008 for the Nino 3.4 region were between 0.1 and -1.8 °C, while the actually observed value was -1.9 °C. In fact a comparison of the two figures shows hardly any continuity between the two sets of curves.

The accuracy of ENSO prediction hinges largely upon how correctly the time of onset of the next El Nino or La Nina episode can be predicted, which in itself is a difficult proposition (Camargo et al 2007, Fedorov et al 2003, Kug et al 2005). These episodes usually commence in the months between April and June, after which forecasting their further evolution over the next year or longer is much easier. The ENSO forecasts made between January and March, before the actual onset of the El Nino or La Nina, are always less accurate than those made later. It is worth recalling that even the strong El Nino of 1997-98, could not be well-anticipated before the signs of the initial onset were observed in the northern hemisphere spring of 1997 (Barnston et al 1999). All ENSO prediction models suffer from this seasonal variation in the forecast skill, a fact that has to be borne in mind while making use of ENSO as a parameter for monsoon forecasts.

While the skill in ENSO forecasts differs widely from one model to another, there is a basic pattern of predictability that is reflected similarly across most of the present models. The predictive skill for forecasts made in March is high for only 2-3 months, while for forecasts made in August the skill extends to longer lead times (Camargo et al 2007). This is another aspect that has an important bearing on the long range forecasts of the monsoon which have to be made in April-May.

### 6.2.4 ENSO and the Northeast Monsoon

This book has addressed itself almost wholly to the Indian southwest monsoon and the methods of predicting the southwest monsoon rainfall. The spatial domain covered by the northeast monsoon is very small compared to the southwest monsoon which is a countrywide phenomenon. The northeast monsoon gives rains to the southern parts of the Indian peninsula and Sri Lanka in the months of October, November and December. The alignment of the orography of southern India is such that it comes within the rain shadow region of the southwest monsoon and has to wait for its share of the annual rainfall until the southwest monsoon has withdrawn from the country and given way to the northeast monsoon. Much of the rainfall of the northeast monsoon is received in association with easterly waves over the Bay of Bengal, and the tropical cyclones forming over the Bay and moving westwards towards Tamil Nadu.

As compared to the southwest monsoon, not much work has been done towards the understanding of the interannual variability of northeast monsoon rainfall or its seasonal prediction. As per Pankaj Kumar et al (2007), there is considerable spatial homogeneity of the northeast monsoon rainfall among the seven meteorological subdivisions in the southern peninsula. The normal area-weighted average rainfall over the northeast monsoon region is 33.8 cm with a standard deviation of 7.3 cm, and it is 28% of the annual rainfall. Previous studies such as those of Raj (2003) and De et al (1999) have indicated that in the case of the northeast monsoon, years of excess rainfall have generally been associated with El Nino years and those of deficient rainfall with La Nina years, which is opposite to what is observed with the southwest monsoon. As against the weakening of the inverse relationship between the southwest monsoon rainfall and ENSO that has been noticed in recent years, Pankaj Kumar et al (2007) have shown that the direct relationship between the northeast rainfall and ENSO has strengthened and became statistically significant after the mid-1970s. This can be attributed to the epochal changes in the regional circulation features. During the recent El Nino years, above normal northeast monsoon rainfall has been received due to stronger easterly wind anomalies and anomalous low level moisture convergence.

### 6.2.5 North Atlantic Oscillation

Walker (1924) while discussing the three oscillations that he had discovered, had commented that India had but little connection with the north Atlantic and north Pacific oscillations, and that for the Indian monsoon it was only the southern oscillation that mattered. This perception has of late been undergoing a change. In several recent research papers about the Indian southwest monsoon, its linkage with the North Atlantic Oscillation is coming into prominence.

In a very recent study, Goswami et al (2006a) have shown that the monsoon does indeed have a link with the North Atlantic Oscillation (NAO). They have found a close relationship between the interdecadal variability of the sea surface temperatures over the North Atlantic Ocean and that of the Indian monsoon rainfall. They have also been able to identify a likely physical mechanism between them, that has so far remained elusive. They defined a summer NAO index as JJAS mean of the difference between normalized sea level pressure anomalies over two $5° \times 5°$ lat/long boxes, one in the north centred around $20°$ W, $65°$ N close to Iceland, and another in the south around $30°$ W, $40°$ N. The southern point was chosen to be over the region where the sea level pressure has the largest variability during northern summer. The summer NAO index was found to have a weak but statistically significant simultaneous correlation with the AISMR.

Sea surface temperatures over the North Atlantic Ocean exhibit an oscillation with a period of about 65-80 years, which is often referred to as the Atlantic Multidecadal Oscillation (AMO). Warm AMO phases occurred during 1860-1890 and 1925-1960 while cool phases occurred during 1895-1925 and 1965-1990. The hypothesis of Goswami et al (2006a) is that a negative AMO produces persistent weakening of the meridional gradient of tropospheric temperature (TT) by setting up negative TT anomaly over Eurasia during northern late summer/autumn resulting in a decrease of southwest monsoon rainfall and its early withdrawal. On the contrary, a positive AMO produces persistent strengthening of the meridional gradient of TT by setting up positive TT anomaly over Eurasia leading to increased monsoon rainfall and a prolonged monsoon season. On interannual time scales, a strong North Atlantic Oscillation (NAO) influences the monsoon by producing similar TT anomaly over Eurasia. Thus there are two complementary teleconnection mechanisms that modulate the interdecadal variability of the Indian monsoon. The tropical ENSO-based mechanism involves a shift of the Walker circulation that influences the regional Hadley circulation, and the extratropical AMO-related mechanism modulates the Indian monsoon through persistent tropospheric temperature anomalies over Eurasia.

This newly found relationship of the monsoon with the north Atlantic Ocean has already found a place in the statistical forecasting model adopted by IMD in 2007. Here, the north Atlantic SST anomaly in December-January over the 20-30°N, 80-100°W domain, has been included as a predictor with a correlation coefficient of -0.45 significant at the 1% level (Rajeevan et al 2006a).

## 6.2.6 Indian Ocean Dipole

The interannual variability in Indian Ocean SST is analogous to that over the tropical Pacific Ocean, and the warming and cooling events over the Indian Ocean reflect the El Nino and La Nina cycle. The atmospheric processes that drive the El Nino produce a similar warming of the Indian Ocean. During the La Nina years, the Indian Ocean is also cooled during winter by the upwelling produced by propagating Rossby waves. However, a new SST-related phenomenon that was peculiar to the Indian Ocean and independent of El Nino and La Nina, was discovered by Saji et al (1999) and Webster et al (1999). It was named as the Indian Ocean Dipole (IOD), as the anomalies in the SST and sea surface height were seen to have opposite signs in the western and eastern parts of the equatorial Indian Ocean. Saji et al defined the Dipole Mode Index (DMI) as the difference in the SST anomaly between two oceanic regions: the tropical western Indian Ocean (50-70° E, 10° S-10° N) and the tropical southeast Indian Ocean (90-110° E, 10° S-equator), shown in Figure 6.2.6.1.

The normal SST pattern over the tropical Indian Ocean shows the eastern region to be warmer than the western region, in association with the westerly winds blowing towards Indonesia. The IOD is said to be in the positive mode when the SST over the southeast equatorial Indian Ocean off Sumatra is anomalously cooler due to unique coastal upwelling and anomalously warmer in the western equatorial Indian Ocean. Cooler waters in the eastern Indian Ocean give rise to easterly winds along the equator. The positive IOD mode usually develops in southern hemispheric winter and matures in spring. The IOD is said to be in the negative mode when the SST over the southeastern region is warmer than normal and the SST over the western region is cooler than normal. In the negative dipole mode, the normal westerly flow gets intensified. Thus the IOD is an internal mode of variability in the Indian Ocean that leads to climate oscillations. The anomalous SST and wind fields associated with the IOD clearly bring out the ocean-atmosphere coupling over the Indian Ocean.

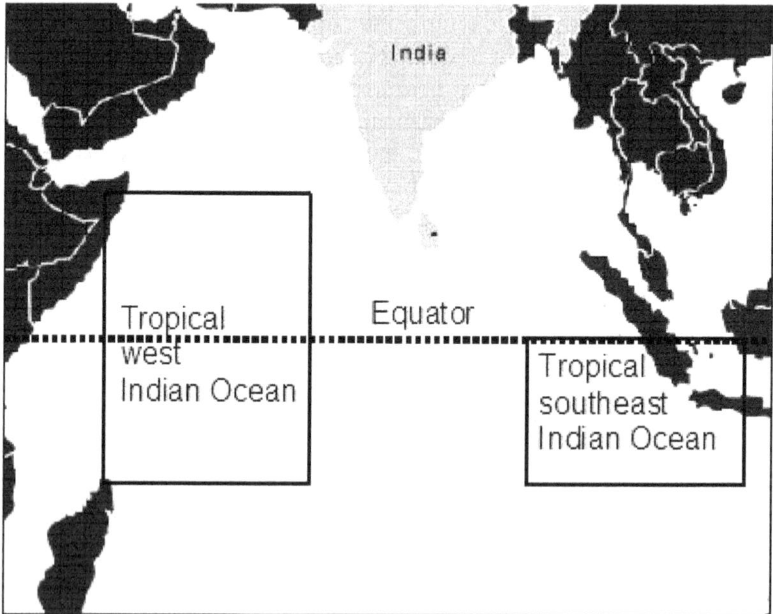


Figure 6.2.6.1 The Indian Ocean Dipole (IOD) regions

Chowdary et al (2007) have observed that the basin-wide warming of the Indian Ocean is stronger in years of concurrent El Nino and IOD. They have discussed the different mechanisms that are responsible for the basin-wide warming when El Nino alone is present and when there is a simultaneous occurrence of El Nino and IOD, and how ocean dynamics play an important role in winter warming of the western Indian Ocean during the IOD years.

While the influence of ENSO events on the Indian monsoon rainfall has been the subject of numerous investigations, there have been comparatively very few studies of the relationship of the Indian monsoon with SSTs over the Indian Ocean. Kothavale et al (2007) have examined the long term trend in SSTs over the Arabian Sea, Bay of Bengal and equatorial South Indian Ocean in the context of global warming. They have also made an attempt to correlate the SST variations in these three ocean areas with the AISMR as well as the monsoon rainfall over smaller homogeneous regions of India. As per their study, there is a positive and statistically significant correlation between SSTs over the Arabian Sea in winter, particularly in December, with the AISMR. After removing the Nino3 effect, the Bay of Bengal and equatorial South Indian Ocean SSTs in spring show a significant and positive correlation with the AISMR and it is even better than that for the Arabian Sea.

Reddy and Salvekar (2003) studied the eastern equatorial Indian Ocean, where anticyclonic twin gyres evolve in the month of May and migrate towards the western part of the Indian Ocean. Their results indicated a negative correlation between June SST over this region and the July-September rainfall over India. Gadgil et al (2004) who compared the OLR patterns of 1986 and 1994 also observed that convection over the western part of the equatorial Indian Ocean is favourable for the monsoon while that over the eastern part is unfavourable.

After the recent discovery of the IOD, there has been considerable interest towards detecting its role in modulating the variability of the Indian monsoon rainfall. Early studies did not, however, find this role to be very dominating. On the contrary, the Dipole Mode Index (DMI) appeared to be highly correlated with the rainfall over eastern Africa and western equatorial Indian Ocean, rather than with the rainfall over the Indian region. Ashok et al (2001) were able to attribute the reduced impact of the El Nino on the Indian monsoon in 1997 to the presence of a positive DMI, but the influence of a negative DMI was not clear enough. A long-period statistical analysis of the IOD mode in relation to the monsoon rainfall was carried out by Kulkarni et al (2007), and a significant result that emerged out of it was that the relationship was stronger following the monsoon than before it. Since the dipole mode peaks in autumn, that is after the monsoon, the influence of the monsoon on the DMI should sound obvious, rather than the monsoon getting influenced by the DMI. The monsoon-DMI relationship also showed a long term trend, having been stronger prior to 1960 and showing signs of weakening thereafter.

Although the IOD and ENSO cycles work independently of each other, both of them have to be taken into consideration in understanding the role of Indian Ocean SSTs on the Indian monsoon. About one-third of the past IOD years have coincided with ENSO events, and two-thirds have occurred in the absence of an El Nino. Gadgil et al (2004) have attempted to synthesize the contributions of ENSO and the convection over the Indian Ocean vis-à-vis the monsoon rainfall. They have shown that there is an interplay of the ENSO and the IOD modes which can lead to different patterns of SST and OLR anomalies over the equatorial Indian Ocean, depending on the phases and strengths of the two modes, with different implications for the Indian Monsoon. The impact of the two modes is additive, and large deficit and excess monsoon rainfall are strongly related to an index that is a composite of the two modes. They have isolated the atmospheric component of the coupled IOD mode and named it as EQUINOO, which is the Equatorial Indian Ocean Oscillation between the positive and negative phases of the IOD mode.

Bhatia et al (2007) in their analysis of the IMD operational long range forecasts for the monsoon rainfall of 2007 have partially attributed the above normal activity over south peninsula and the quick revival of the monsoon in the second week of September to the positive phase of the Equatorial Indian Ocean Oscillation (EQUINOO) defined by Gadgil et al (2004). The positive phase of EQUINOO is favourable for the monsoon.

That the IOD and ENSO modes need to be synthesized has also been brought out in the recent work of Krishnamurthy and Shukla (2007) on the space-time evolution of convection over the monsoon region covering the Indian subcontinent, the Indian Ocean and the western Pacific Ocean. A highlight of this work is the multi-channel singular spectrum analysis of the daily OLR field, which has yielded two intraseasonal oscillatory patterns and two large scale standing patterns as the most dominant modes of intraseasonal variability of the monsoon. The two oscillatory modes have time scales of 45 and 28 days and their average cycles of variability correspond to the active-break cycles. Krishnamurthy et al showed that during an active (break) cycle, a convection (dry) anomaly zone develops in the equatorial Indian Ocean, then covers the Indian subcontinent and finally dissipates over the north. One of the two standing modes has anomalies of uniform sign covering the entire region and is ENSO-related. The other standing mode has a dipole structure in the equatorial Indian Ocean associated with large scale anomalies over

India with the same sign as those over the western part of the dipole. Both these standing modes were seen to persist throughout the monsoon season, maintaining their individual identities. The seasonal mean monsoon was mainly determined by the two standing patterns, and not much by the oscillatory modes.

The relative importance of IOD and ENSO modes to the monsoon has been the subject of yet another study by Ashok et al (2006) in which they have considered both all-India and subregional scale rainfall. They have shown that the IOD events have a significant impact on rainfall in the monsoon trough region, southwest coastal region of India, and also over Pakistan, Afghanistan, and Iran. ENSO events have a wider and stronger impact but it is opposite to that of IOD events.

## 6.3 Advances in Dynamical Models

In recent years there have been very significant advances in the design and performance of dynamical models. At the same time, it is very important to note that many aspects of dynamical models have remained inherently problematic and satisfactory solutions have yet to be found to eliminate them. The problems in fact get further accentuated when the models are applied to the tropical regions which are essentially the monsoon regions. The geostrophic approximation does not hold good in the near-equatorial regions of the tropics. Over vast areas of the tropics, conventional observations are not available for the purposes of data initialization on the scale that the models demand. Assimilation of satellite data as a proxy for in situ observations has its own problems. By far, the greatest problem arises from the fact that convective clouds dominate tropical weather, and the cloud-radiation feedback and the interaction with the land and sea surface and the boundary layer, is very complex. Most existing models are not robust enough to parameterize them realistically and many of the interactions have in fact not even been fully understood. All these factors continue to put serious limitations to monsoon prediction in an operational mode and call for more research effort to overcome them. Krishnanurti (2004) has identified monsoon precipitation as one of the central scientific problems in this area.

On the positive side, a variety of conventional and satellite-derived data is now available to the models, the accuracy of the satellite products has improved and the models are now more capable of assimilating them. Immense computational power is now available for running both global and regional models at a much higher resolution and with a finer representation of orography than before. The models are now capable of incorporating the physical parameterizations of convection, radiative transfer, boundary layer physics and land surface processes in a much more realistic manner. However, all this has not resulted in a corresponding visible improvement in the skill of dynamical monsoon prediction, in respect of both intraseasonal and interannual variability.

*NCMRWF:* The National Centre for Medium Range Weather Forecasting (NCMRWF) is the only organization in India where global data assimilation and medium range weather forecasting is done operationally. This centre was established in 1989 with the implementation of the ECMWF (cycle-30) global model. In 1993, the NCEP model T80L18 was implemented and this model is being run operationally since then. In 2007, NCMRWF implemented a new Global Forecast System at T254L64 resolution on the IBM P5 cluster-based Param Padma and Cray-X1E computer systems. The new GFS and the data assimilation system has been adopted from NCEP.

*Mesoscale Models:* The usefulness of mesoscale models has been amply demonstrated in several case studies of tropical cyclones, western disturbances and active monsoon spells over India. Since 2002, NCMRWF has been using the MM5 model to generate real time forecasts for the western Himalayan region. Many other Indian research institutions have also been carrying out specific studies using mesoscale models, the MM5 model being the one most popularly used. The WRF model is also coming into use in a real time test mode in Indian operational agencies and it is showing some potential. However, Mohanty et al (2003) have made several important points regarding the use of mesoscale models. First, the physical processes have to be parameterized or represented explicitly depending upon their nature and the resolution of the model. Second, the information regarding the characteristics of the land surface and orography has to be prescribed in such detail as demanded by the high resolution of the model. Third, there is a need to organize field experiments to generate mesoscale observations that would help to validate the model forecasts and to improve the parameterization schemes themselves.

*Superensembles:* The multi-model superensemble approach to subseasonal and seasonal forecasting has shown particular promise with regard to monsoon rainfall prediction. The philosophy of the superensemble which has been pioneered by Krishnamurti et al (2000) and is being constantly refined by them (Chakraborty et al 2006, Krishnamurtii et al 2007), is to analyze the past behaviour and biases of the models that form the ensemble and apply this knowledge to get a better forecast. The methodology thus involves a training phase and a forecast phase. The multiple regression statistics between the model anomalies and the analysis fields reveal the areas of major skills of the member-models. The superensemble can then be made to bring out the best out of each member-model to arrive what could be appropriately termed the collective wisdom of the superensemble. However, this is not as simple as it may appear, and the amount of statistical data required to be processed could be staggering. If the ensemble consists of 10 member-models providing forecasts over an array of roughly 100,000 locations at 10 vertical levels for 10 variables, then 10 million correction weights would need to be applied!

In the conventional superensemble, the training phase varies from 120 days for medium range forecasts, to 10 years for seasonal climate forecasts. The correction weights assigned to the models could be less than 1 or even negative, depending on the past behaviour of the models. To improve further upon the conventional superensemble method, additional synthetic data sets are constructed from the member-model forecast data using a combination of the past observations and past forecasts. A consistent spatial pattern is determined among the observations and forecasts using a linear regression relationship in the EOF space. Sets of such synthetic forecasts are then obtained for each available member model forecast and used for the creation of superensemble forecasts.

The superensemble method has indeed a promising future, but currently it is not being used widely because of the immense computational resources that it demands.

*Climate Models:* While making weather predictions several days ahead using dynamical models, the accurate specification of the initial conditions of the atmosphere is a prime necessity. Climate models, however, do not take into consideration the evolution of weather systems on such a short time scale and so it is not the initial conditions but the parameterization of radiative forcings and ocean processes that decides their accuracy. Climate models are steadily improving in these respects and some models treat aerosol-related radiative forcing interactively, but most models maintain a stable climate without flux adjustment.

The dynamical models that are currently in use differ widely with regard to the complexity of their structure and the manner in which they incorporate the physical processes of the atmosphere-land-ocean system (Solomon et al 2007a). The simple climate Models (SCMs) are built around an energy balance equation, a prescribed value of climate sensitivity and a basic representation of ocean heat uptake. These simple models are able to provide us with broad estimates of the increase in global mean temperature and the rise in sea level due to thermal expansion. SCMs can also be coupled to simplified models of biogeochemical cycles to find out the climate response to a wide range of emission scenarios.

Earth System Models of Intermediate Complexity (EMICs) include some dynamics and parameterizations of the atmospheric and oceanic circulations, and biogeochemical cycles. The possible long term changes in climate can be estimated by executing the model runs over several centuries and in large ensembles, but this is usually done at the cost of spatial resolution, so only continental scale results can be obtained.

Both SCMs and EMICs are computationally not very demanding. So they can be made to generate probabilistic climate projections for the wide range of emission scenarios that have been envisaged and to evaluate the sensitivity of the climate to possible changes in various parameters. They are also useful for tuning the more advanced models.

The AOGCMs incorporate the dynamics of the atmosphere and oceans and land surface processes, and include sea ice and other components. AOGCMs are currently being run by several centres around the world and over 20 AOGCMs have been participating in IPCC AR4 related experiments. While climate projections made by different models generally tend to agree qualitatively, they most often differ in many respects, especially when there is an attempt to quantify the results. Although the dynamics of AOGCMs are becoming increasingly comprehensive, the uncertainties in the parameterization of physical processes are the principal cause of the differences between model results. Another cause is that the lower resolutions of some of the models do not allow the finer scale features to be captured by them. Therefore, when it comes to taking practical action, it becomes imperative to judge how much the model results are to be believed, which results could be assigned a higher credibility than others, and what choice to make when the results are conflicting. For this purpose, many international efforts have been launched to compare the performance of different models in experiments performed under common conditions. Since the Atmosphere-Ocean General Circulation Models (AOGCMs) are being used as the primary tool for simulating past climates and making future climate projections, it is important to apply norms for evaluating the models and interpreting their results (Solomon et al 2007b).

A crucial issue in the acceptance of climate model projections is to know how realistically the model is capable of simulating the mean climate as we presently know it. If a climate model cannot pass this benchmark test, it goes down in the level of credibility that can be assigned to its future projections. Here again, there is no single climate model that can faithfully simulate the present climate in every respect. For example, some models may be good in simulating the monsoon rainfall pattern on a spatial or temporal scale in a relative manner but not the rainfall amounts in absolute terms. Then there could be some models that correctly simulate the monsoon seasonal rainfall

(JJAS total) but fail in the distribution over each of the four months. It is necessary to evaluate the strengths and weaknesses of various models with regard to such details which may not be important on a global scale, but could be crucial on a regional or local scale.

In recent years, models simulations of the present climate have undoubtedly improved in general, but compared to sea level pressure and surface temperature, deficiencies still remain in the simulation of tropical precipitation. As regards the simulation of climate variability, most climate models are now capable of simulating the dominant modes of climate variability observed over the extratropics such NAM, SAM, PNA and PDO but many models still have problems simulating ENSO and most models cannot simulate MJO properly.

Another test of model performance is the ability to simulate past climates, which is more difficult to prove because of the non-availability of the required data for past periods. However, if a model can provide a physically self-consistent explanation of past climate variations then it can justifiably be relied upon for its estimates of the future climate.

A few climate models have been shown to have the capability of making predictions on the seasonal and subseasonal scale when the initial conditions are properly prescribed. Such studies also add to the confidence that they are adequately representing some key processes and teleconnections in the climate system.

*Interpretation of Model Results:* While interpreting the results of any dynamical model, it is of vital importance to know what type of model it is, its limitations, past history, the areas of its strengths and weaknesses and so on. Crude models would obviously produce crude results. Sophisticated and advanced models might bring out conflicting features in their prediction, some acceptable and some not. Particularly in the case of climate models, it is absolutely necessary to have an assessment of the model done before the results are acted upon. Models that lead to scary scenarios like the metropolis of Mumbai getting submerged by 2100, or the monsoon drying up by 2060 and the like, should be first investigated thoroughly to find out what is the credibility of such predictions.

Even on the seasonal and subseasonal scales of the monsoon rainfall, there are some models which perform better in terms of the rainfall pattern, while some other models may perform better in terms of the rainfall amounts. There may be models which are good at predicting only droughts. Although the merits and demerits of the models can be compared in hindsight, the problem arises as to which model to believe out of the several alternatives when it comes to real time prediction. Diagnostic studies are no doubt important but they do not help much for purposes of prognostication.

*Seasonal Prediction of the Indian Monsoon:* An intercomparison of AGCM model runs for simulating the Indian monsoon was carried out ten years ago (Gadgil et al 1998). A fresh assessment is currently under way of the manner in which AGCMs are able to make a seasonal prediction of the Indian monsoon under the multi-institutional SPIM project. Several scientists and modelling groups of the country have ported their models on the computational platform of the Param Padma Terraflop facility of CDAC, Pune. The objective is to generate hindcasts for a period of 20 years 1985-2004 with observed SST as the boundary condition, and with initial conditions drawn from NCEP reanalysis from 26 to 30 April as 5-member ensemble. The model outputs are to be compared with the observations. For five selected extreme monsoon years, 1987, 1988, 1994, 1997 (El Nino) and 2002, runs have being made assuming that the SST anomalies persist from April onwards.

The results of SPIM will ultimately help in deciding which of the models used in the intercomparison fare the best in comparison with the observations. But early results indicate that there is no model which stands out uniquely over others in an overall manner. Some models are good at predicting droughts, some at excess rainfall years. The prognoses are all model-dependent and conflicting.

*India's Own Models?:* While models developed abroad are being operationally used by IMD and NCMRWF, and several Indian scientists have been using dynamical models for making experiments over the Indian region and for monsoon studies, there has hardly been any effort towards constructing a dynamical model of our own. The development of an Indian model is required because it would focus on India's special need for the prediction of monsoon rainfall on all scales: intraseasonal, seasonal, interannual, decadal and centennial. We just cannot afford to take crucial decisions on the basis of imported models, whether it is on the short range with regard to disaster management or with regard to global warming on the climate scale. We need to have our own models because the people who develop them would have a greater sense of responsibility to the country, and the people who have to act upon them can do so with a greater degree of confidence.

## 6.4 Monsoon Field Experiments

Unlike other physical sciences in which most major advances have resulted from laboratory experiments, significant breakthroughs in atmospheric and ocean sciences have come out of specially organized in situ observations. The Arabian Sea low level jet associated with the southwest monsoon, the formation of the monsoon onset vortex, the development and collapse of the Arabian Sea warm pool prior to the monsoon onset, have all been discovered during the special field experiments undertaken in this area. Meteorological and oceanographic observations are important for purposes of synoptic scale monitoring, but they lack the detail or coverage required for obtaining an insight into the several physical processes that need to be known. Measurements of fluxes at the air-sea interface, for example, cannot be made by satellites or by any other means except by actual observations over the oceans.

### 6.4.1 International Experiments and Programmes

Realizing that there was no alternative to field experiments, the international meteorological and oceanographic communities came together and joined hands with their Indian counterparts in organizing several field experiments over the Indian Ocean region in the 1960s and 1970s. These were: the International Indian Ocean Experiment (IIOE 1961-1964), Indo-Soviet Monsoon Experiment (ISMEX 1973), the International Monsoon Experiment (MONSOON 1977) and the Monsoon Experiment (MONEX 1979).

IIOE was the first attempt at carrying out integrated atmospheric and oceanographic investigations of the Indian Ocean about which very little was known earlier from the scientific point of view. The observations made during the IIOE revealed the presence of the low level jet over Somalia and the Arabian Sea during the southwest monsoon, the strong atmospheric inversions that prevail over the Arabian Sea and the moisture transport that occurs from the equator into the Indian landmass. ISMEX-73 was a limited bilateral exercise and MONSOON-77 was a prelude to MONEX-79.

MONEX-79 was by all standards a landmark effort involving the participation of 20 ships, and several aircrafts from which dropsondes were released to obtain atmospheric profiles over the ocean. The comprehensive MONEX data set that was generated has served as the basis for a very large number of national and international investigations related to the southwest monsoon.

Subsequent to MONEX-79, however, there was a generally lean period of field activity, and the next international field experiment in this area, the Joint Air-Sea Monsoon Interaction Experiment (JASMINE) was conducted in 1999 over the tropical Indian Ocean during the premonsoon months and towards the end of monsoon season.

The Indian Ocean Experiment (INDOEX) was organized in February-March 1999 over an area covering 30° N to 30° S, and 50° E to 100° E. The objective was to monitor how the northeasterly winds carried the low level outflow of pollutants from India into the tropical convergence zone, where cloud processing would modify the properties of the aerosols. During INDOEX, measurements were made of the haze that spreads at this time of the year over north Indian Ocean and south and southeast Asia, and its contribution to regional climate forcing was assessed. In both JASMINE and INDOEX experiments, Indian scientists had a rather secondary involvement.

In recent years, the focus of international experiments and observational programmes has shifted from the Indian southwest monsoon in particular to the Asian monsoon in general, and the African and American monsoons have also become a subject of attention. As far as the Asian monsoon is concerned, the international initiative is now coming from mainly three countries, Japan, China and Korea. The other shift of emphasis is that these new programmes have major hydrology and water cycle related components built into the effort.

The GEWEX Asian Monsoon Experiment (GAME) was carried out under the overall umbrella of the Global Energy and Water Cycle Experiment between 1996 and 2005. The objectives of GAME were: (1) to understand the role of the Asian monsoon in the global energy and water cycle, (2) to improve the simulation and seasonal prediction of the Asian monsoon, and (3) to assess the impact of monsoon variability on the regional hydrological cycle. Japan, China and Korea were the main players in GAME. Special observation programmes were conducted over four selected regions: Siberian Taiga forests and Tundra area, the Tibetan plateau, the Huai-he river basin in China, and the Chao Phraya river basin in Thailand. The last phase of GAME was devoted to a detailed analysis of the observations and modelling studies.

The interaction between atmospheric aerosols and the Asian monsoon is an aspect that has currently been gaining importance. It is speculated that in the summer monsoon season, fine aerosols could get accumulated at the Himalayan foothills and the Tibetan plateau. They could act as a high-altitude heat source and enhance the rainfall in June, but by cooling the land surface they may suppress the rainfall in the months of July to September.

The deposition of dust on the snow surface could reduce its albedo and hasten the retreat of Himalayan glaciers.

The impact of elevated aerosols on the radiation-monsoon-water cycle interaction was the theme of a conference held in Xi'ning, China, in August 2006. The need to investigate this and other aspects of the Asian monsoon was discussed at this conference and the concept of a new international experiment called the "Asian Monsoon Year 2008 (AMY08)" was mooted. It has since been accorded a formal status by the international community and Japan and China are playing a major role in its planning and execution.

The "Asian Monsoon Year (AMY08)" experiment of 2008-2009 is a crosscutting observation and modelling effort aimed at understanding the radiation-monsoon-water cycle interaction and the ocean-land-atmosphere interaction of the Asian monsoon system, and on improving monsoon prediction. This goal will be reached through integration of the ongoing and planned national projects in Asian countries and other international research programmes. A wide range of scientific questions are being addressed by AMY08 (MAIRS 2007):

Monsoon modelling: What determines the structure and dynamics of the diurnal and annual cycles of the coupled atmosphere-ocean-land system? What are the major weaknesses of the climate models in the simulation of these cycles? Will removing these weaknesses improve the modelling of intraseasonal to interannual variability?

Interannual monsoon rediction/predictability: What is the current accuracy of dynamical monsoon seasonal predictions and how to improve it? How predictable is the monsoon interannual variability?

Monsoon intraseasonal oscillation: What are the critical processes causing the intraseasonal and seasonal variation and what are the roles of multiscale interaction? To what extent are the intraseasonal variations predictable? What are the major challenges to modelling and predicting such oscillations and the MJO and the monsoon intraseasonal variations?

Interdecadal variation and anthropogenic climate change: What are the major modes of interdecadal variation of the monsoon system? How will the monsoon system change in a global warming environment? What are the subseasonal to interannual factors that influence extreme events? What is the sensitivity of the monsoon to external and anthropogenic climate forcing?

A wider programme named the "International Monsoon Year (IMY)" has been mooted to cover other monsoon activities.

The Monsoon Asian Hydro-Atmosphere Scientific Research and Prediction Initiative (MAHASRI) is a very new initiative the objective of which is to develop a hydro-meteorological prediction system, particularly with the time scale up to a season, through a better scientific understanding of the variability of the Asian monsoon. The planning for MAHASRI began in August 2005 and it was conceived as a successor to the GEWEX Asian Monsoon Experiment (GAME). MAHASRI is primarily supported by Japan and the Japan National Committee for MAHASRI was constituted in July 2006. However, many other countries like China, Thailand, Vietnam, and Bangladesh have joined the programme or are considering joining it. A more concrete collaboration with Asian hydrometeorological agencies and research institutes, especially in southeast Asia is expected to be achieved.

While MAHASRI is essentially a continuation of the hydrometeorological component of GAME, it has some wider objectives. An expansion of the target scientific field from air-land interactions in GAME to air-land-sea interactions, is envisaged with a closer collaboration with the CLIVAR community. The area is also to be expanded beyond the GAME area to cover the maritime Continent, Western Pacific Ocean and Indian Ocean and India, while it may no longer include Siberia. MAHASRI will include investigations into the winter monsoon.

MAHASRI is expected to establish a prototype system that would demonstrate the recording of hydrometeorological observations and data transmission in real time, which will help in flood management in Asian monsoon countries. It will also generate weekly to monthly scale dynamic probalistic monsoon predictions for water resources and agricultural management in these countries. For this purpose, intense observational programmes will be carried out for limited periods.

The observational phase of MAHASRI is planned to coincide with the Asian Monsoon Year 2008 and its research phase with the International Monsoon Year in 2011.


## 6.4.2 Indian Climate Research Programme

It is understandable that the level of international interest in the Indian southwest monsoon and the degree of participation in international programmes related to the monsoon would vary with time, and depend upon several factors like availability of resources and observational tools, the focus of the investigations, and even political expediency. India, however, has an enduring interest in developing a better understanding and prediction capability of the monsoon rainfall on which the entire Indian population and the Indian economy are dependent. Hence, while the participation of Indian scientists in international monsoon related programmes is desirable and would even be beneficial, it is necessary that they do not lose sight of the country's own interests and priorities

It was basically with this thinking that the Indian Climate Research Programme was mooted in 1996 and it is currently under implementation (DST 1996, 1998). The distinctive feature of ICRP is that it is an endeavour that cuts across several government ministries, national institutions and universities and can thus bring together a pool of multi-disciplinary talent that exists in the country and optimize the use of observational infrastructure. The main objective of ICRP is to achieve a deeper understanding of the physical processes that govern the atmosphere-ocean-land system and the space-time variation of the monsoons on various scales ranging from subseasonal to decadal. ICRP also aims at making a realistic assessment of climate change and climate variability on longer time scales, and their impacts on critical resources and India's agricultural productivity.

### 6.4.3 BOBMEX, ARMEX and CTCZ

The first field experiment to be organized under ICRP was the Bay of Bengal Monsoon Experiment (BOBMEX) in July-August 1999 which was preceded by a pilot experiment in October-November 1998. BOBMEX was designed to critically study the air-sea coupling during the monsoon season and its variability with a prime focus on the Bay of Bengal. Special intensive observation periods were organized during the experiment, in which data from ships, satellites, radars, and ocean buoys, were pooled together with ground-based and upper air observations. Although the activity of the 1999 monsoon was generally weak, valuable field observations were collected in respect of intraseasonal oscillation of organized convection, northward propagation of cloud bands, and active-weak cycles of the monsoon.

The Arabian Sea Monsoon Experiment (ARMEX) was the second field experiment carried out under the umbrella of the ICRP. ARMEX had two major objectives. One was to study the evolution, maintenance and collapse of the Arabian Sea warm pool in the pre-onset phase of the monsoon. The other was to study the Arabian Sea convection associated with intense rainfall events on the west coast of India during the monsoon and particularly the off-shore trough. ARMEX was executed in two separate phases to address these different issues related to the atmosphere and ocean: June-August 2002 and March-June 2003, and some of the measurements were completed later in 2005. As it happened, July 2002 was a month of unprecedented rainfall deficiency over India and the arrival of the 2003 monsoon over Kerala was marked by a week's delay and the absence of an intense convective system.

Several observational platforms were deployed during ARMEX, such as met-ocean buoys and a micrometeorological tower at Goa. Intensive XBT surveys were carried out and special aircraft missions were undertaken to investigate aerosol-cloud microphysical properties over the Arabian Sea. As many as 200 scientists from 25 different organizations in the country participated in ARMEX.

The achievements of BOBMEX and ARMEX and the new scientific results that have emerged out of these experiments have been summarized by Sanjeeva Rao and Sikka (2005), Bhat et al (2001), and Bhat and Narasimha (2007).

The next programme in the continuing series of ICRP field experiments is the Continental Trough Convergence Zone (CTCZ) experiment which is planned to be conducted in the 2008-2010 time phase. While the primary focus of this experiment would be on the area of the Indo-Gangetic plains, this is a long term and highly ambitious programme which envisages an investigation of the monsoon as a total land-atmosphere-ocean system. The experiment is designed around different types of observational platforms that would enable large scale, mesoscale and sub-basin scale processes to be studied (Sanjeeva Rao and Sikka 2007).

## 6.5 Future Indian Satellites for Weather and Climate

Since 1982, the Indian Space Research Organisation has launched a series of geostationary satellites that have provided a continuous meteorological coverage of the Indian region and the surrounding land and Indian Ocean regions. Four satellites in the INSAT-1 series, three in the INSAT-2 series, the dedicated Kalpana-1 satellite, and the current INSAT-3A satellite, have carried a total of nine Very High Resolutions Radiometer (VHRR) instruments so far, besides the Charged Coupled Camera (CCD) devices on the more recent satellites. The next satellite, INSAT-3D, to be launched in 2008 will have an advanced 6-channel imager and a 19-channel sounder. These will offer new capabilities for obtaining high resolution images in new channels and deriving vertical profiles of temperature and moisture.

India has an ambitious space programme that includes a lunar mission. While the accent on the INSAT programme has always been on meeting the requirements of operational meteorology, India is now planning to go in a big way in several new directions towards making systematic observations of parameters related to climate studies. This will also involve the use of microwave sensors, both passive and active, and the placement of satellites in non-geostationary orbits.

Geostationary and polar orbiting satellites have their own relative merits and demerits in various aspects. Geostationary satellites provide images every half-hour or even faster for small sectors. However, because of their height, they are not suited for microwave remote sensing as the radiance reaching them is very weak. Orbiting satellites either in polar orbits or tropical orbits like TRMM, can provide a higher spatial resolution but they have a repeat cycle of one or two days and hence many atmospheric developments are lost between scans. The proposed Megha-Tropiques satellite strikes a trade-off among these conflicting requirements by placing passive microwave sensors in a low altitude orbit also having a low inclination with respect to the equatorial plane.

Megha-Tropiques is a joint India-France (ISRO-CNES) mission with a shared responsibility for development of payloads which would be flown on an ISRO IRS bus. A PSLV launcher will launch the satellite from Sriharikota in an orbit with 867 km altitude and a unique $20°$ inclination in 2009. The expected mission life is 3 years.

The main scientific objectives of the Megha-Tropiques mission are:

(a) To collect a long term set of measurements with a good sampling and coverage over tropical latitudes to understand better the processes related to tropical convective systems and their life cycle.
(b) To improve the determination of atmospheric energy and water budget in the tropical region on various time and space scales.
(c) To study tropical weather and climate events like monsoon variability, droughts, floods, and tropical cyclones, and their predictabilty.

Megha-Tropiques will carry a rare combination of three state-of-art payloads, MADRAS, SAPHIR and ScaRaB designed for measurements of radiative fluxes, precipitation, humidity profiles and cloud properties, which are described below.

MADRAS (Microwave Analysis and Detection of Rain and Atmospheric Structures), will be a passive imaging radiometer operating at five frequencies of 18.7, 23.8, 36.5, 89 and 157 GHz in both H and V polarizations except the 23.8 GHz which will have only V polarization. Data from the first three channels will have applications in the retrieval of rain over oceanic regions, liquid water content in clouds and vertical integrated water vapour. Their spatial resolution is expected to be better than 40 km. The 89 GHz channel will be useful in retrieving convective rainfall over both land and ocean at a still better resolution of less than 10 km. The 157 GHz channel will measure the concentration of ice particles in clouds at a resolution as high as 6 km.

SAPHIR (Sounder for Atmospheric Profiling of Humidity in the Inter-tropics by Radiometry) is a microwave sounding instrument. It will have six channels in the frequency region of 183 GHz, having 10 km ground resolution. SAPHIR soundings will complement the temperature and humidity profiles that will be derived from the INSAT-3D sounder that is to be launched in 2008.

The third Megha-Tropiques payload, is a radiation budget instrument. ScaRaB (Scanner for Radiation Budget Measurement) will have four channels: Sc1 - Visible 0.5 to 0.7 $\mu$, Sc2 - Solar 0.2 to 4.0 $\mu$, Sc3 - Total 0.2 to 100 $\mu$, and Sc4 - IR window 10.5 to 12.5 $\mu$. Sc2 and Sc3 are the main channels of the ScaRaB instrument. Longwave irradiance can be calculated from the difference between Sc3 and Sc2 measurements. Images from Sc1 and Sc4 channels will be used for scene identification and will provide the necessary compatibility with operational satellites like INSAT which have radiometers with similar spectral channels.

The time frame of the Megha-Tropiques mission coincides with that of the forthcoming Indian satellite, INSAT-3D which will have an advanced imager and a sounder, and another Indian satellite, Oceansat-2, which will fly a scatterometer for measurement of surface winds. A synergistic uitlisation of the data gathered from all these satellites is certainly going to lead to a great advancement of the current knowledge of the role of the tropical atmosphere and oceans in the global weather and climate.

The progress of the Megha-Tropiques mission can be seen on the web sites http://meghatropiques.ipsl.polytechnique.fr/ and http://www.isro.gov.in/.

These are all exciting developments and the Indian weather and climate community is looking forward to their realization.


## 6.6 How Predictable is the Monsoon?

It is logical to expect that with the evolution of more credible statistical and dynamical models, a better understanding of the changing global relationships of the monsoon, the unravelling of hitherto unknown monsoon processes through field experiments, the increased availability of satellite-based data and its improved assimilation into dynamical models, monsoon prediction will only get better on all scales. However, this optimism is not shared by many, and there is a line of thought which suggests the there is an intrinsic limitation to the predictability and whatever efforts may be put in, this predictability barrier cannot be crossed. This would mean that even if the current deficiencies, particularly those of the dynamical models, are eliminated, the goal of accurate monsoon prediction would remain unattainable.

AjayaMohan et al (2003) have proposed a way of defining the potential predictability of the monsoon quantitatively as the ratio of the external and internal interannual variances. On the basis of an analysis of 55 years of daily circulation data, Goswami (2004) showed that the external low frequency interannual forcing such as ENSO is weaker than the internal low frequency variability generated by the vigorous intraseasonal oscillations. As a result the potential predictability of the monsoon generally remains low and it is further influenced adversely by the interdecadal oscillation of the monsoon. According to Goswami, the potential predictability of the monsoon has reduced by half from the period 1951-1970 to the period 1981-2000. This decrease is due to a much larger reduction of the external variability compared to only a marginal reduction in the internal variability.

In a more recent study, Goswami et al (2006b) have re-addressed the issue of the predictability of the monsoon. Their presumption is that the monsoon intraseasonal oscillations (ISOs) cannot be isolated from the monsoon annual cycle (MAC) and both are an integral part of the monsoon system. The ISOs even play an important role in defining the climatological mean MAC. Hence for a realistic simulation of the mean monsoon rainfall, the model should be able to simulate both the observed external component of the interannual variability and the ISOs. This ideally demands an atmosphere-ocean coupled model, but even AOGCMs have their own biases that could come in the way.

Accordingly to Kang and Shukla (2005), there are three essential requirements for a successful model prediction of monsoon rainfall: a large and persistent anomaly at the earth's surface, a well-defined dynamical response of the monsoon rainfall to the change in boundary condition, and a large signal-to-noise ratio for this response. As all the three requirements are yet to be met by any single dynamical model, it has not been possible to realize the potential predictability of monsoon rainfall.

Many intercomparisons of atmosphere-only GCMs (Section 4.12) have clearly brought out their systematic errors in simulating the mean monsoon rainfall pattern and its variability. The involvement of the ocean therefore seems to be inevitable as the next logical step towards improving the simulations. However, as far as the monsoon is concerned, contrary to what happens elsewhere, the SST anomalies in the Indian and Pacific Oceans are themselves forced by the atmosphere or they have a strong coupling with the atmosphere (Wang et al 2005).

India and the world have come a long way since the times of Edmund Halley who attempted the first scientific explanation of the Indian monsoon, or from the times of Gilbert Walker who discovered the global connections of the monsoon and unravelled many of its mysteries. However, there is long way to go. Monsoon prediction continues to be a challenging and demanding task, but it is worth the effort and it is certainly not a futile one. The time has now come for Indian scientists to begin working towards what monsoon prediction really calls for: an all-encompassing land-atmosphere-ocean coupled model. Considering what we possess today in terms of science and technology, this is a realizable goal.

## 6.7 In Conclusion

The Bible gives an account (Matthew 16:1-4, Luke 12:54-56) of how once people had come to Jesus and asked him to show them a sign from heaven. He had replied,

"When evening comes, you say, 'It will be fair weather, for the sky is red', and in the morning, 'Today it will be stormy, for the sky is red and overcast'.

"When you see a cloud rising in the west, immediately you say, 'It's going to rain', and it does. And when the south wind blows, you say, 'It's going to be hot', and it is.

"You know how to interpret the appearance of the earth and the sky, but you cannot interpret the signs of the times."….

Nature does allow us to predict the future. The monsoon indeed shows its signs. We must only learn to interpret them.

## 6.8 References

AjayaMohan, R. S. and Goswami B. N., 2003, "Potential predictability of the Asian summer monsoon on monthly and seasonal time scales", *Meteorology Atmospheric Physics*, doi:10.1007/s00703-002.

Ashok K., Guan Z. and Yamagata T., 2001, "Impact of the Indian Ocean dipole on the relationship between the Indian monsoon rainfall and ENSO", *Geophysical Research Letters*, 28, 4499-4502.

Ashok K. and Saji N. H., 2006, "On the impacts of ENSO and Indian Ocean dipole events on sub-regional Indian summer monsoon rainfall", *Natural Hazards*, 42, 273-285.

Banerji S. K., 1950, "Methods of forecasting monsoon and winter rainfalls in India". *Indian J. Meteorology Geophysics*, 1, 4-14.

Barnston A. G., Glantz M. H. and He Y., 1999, "Predictive skill of statistical and dynamical climate models in SST forecasts during the 1997-98 El Nino episode and the 1998 La Nina onset", *Bulletin American Meteorological Society*, 80, 217-243.

Bhat G. S. and coauthors, 2001, "BOBMEX – The Bay of Bengal Monsoon Experiment", *Bulletin American Meteorological Society*, 82, 2217-2243.

Bhat G. S. and Narasimha R., 2007, "Indian summer monsoon experiments", *Current Science*, 93, 153-164.

Bhatia R. C., Rajeevan M. and Pai D. S., 2007, "An analysis of the operational long range forecasts of 2007 southwest monsoon rainfall", *NCC Research Report No. 5*, India Meteorological Department, Pune, 35 pp.

Camargo S. J., Barnston A. G., Klotzbach P. J. and Landsea C. W., 2007, "Seasonal tropical cyclone forecasts", *WMO Bulletin*, 56, 297-309.

Chakraborty A. and Krishnamurti T. N., 2006, "Improved seasonal climate forecasts of the south Asian summer monsoon using a suite of 13 coupled ocean-atmosphere models", *Monthly Weather Review,* 134, 1697-1723.

Chowdary J. S. and Gnanaseelan C., 2007, "Basin-wide warming of the Indian Ocean during El Nino and Indian Ocean dipole years", *International. J. Climatology,* 1421-1438.

De U. S. and Mukhopadhay R. K., 1999, "The effect of ENSO/Anti ENSO on northeast monsoon rainfall", *Mausam*, 50, 343-354.

Delsole T. and Shukla J., 2002, "Linear prediction of Indian monsoon rainfall", *J. Climate*, 15, 3645-3658.

DST, 1996, *Indian Climate Research Programme Science Plan*, Department of Science and Technology, New Delhi.

DST, 1998, *Indian Climate Research Programme Implementation Plan*, Department of Science and Technology, New Delhi.

Fedorov A. V. and co-authors, 2003, "How predictable is El Nino?", *Bulletin American Meteorological Society*, 84, 911-919.

Gadgil S. and Sajani S., 1998, "Monsoon precipitation in the AMIP runs", *Climate Dynamics*, 14, 659- 689.

Gadgil S., Vinayachandran P. N., Francis P. A. and Gadgil S., 2004, "Extremes of the Indian summer monsoon rainfall, ENSO and equatorial Indian Ocean oscillation", *Geophysical Research Letters*, 31, doi:10.1029/2004GL019733.

Goswami B. N., 2004, "Interdecadal change in potential predictability of the Indian summer monsoon", *Geophysical Research Letters*, 31, doi:10.1029/2004GL020337.

Goswami, B. N., Madhusoodanan M. S., Neema C. P. and Sengupta D., 2006a, "A physical mechanism for North Atlantic SST influence on the Indian summer monsoon", *Geophysical Research Letters*, 33, doi:10.1029/2005GL024803.

Goswami B. N., Wu G. and Yasunari T., 2006b, "The annual cycle, intraseasonal oscillations, and roadblock to seasonal predictability of the Asian summer monsoon", *J. Climate*, 19, 5078-5099.

Gowariker V., Thapliyal V., Sarker R. P., Mandal G. S. and Sikka D. R., 1989, "Parametric and power regression models: New approach to long range forecasting of monsoon rainfall in India", *Mausam*, 40, 115-122.

Gowariker V., Thapliyal V., Kulshrestha S. M., Mandal G. S., Sen Roy N. and Sikka D. R., 1991, "A power regression model for long range forecast of southwest monsoon rainfall over India", *Mausam*, 42, 125-130.

Gowariker V., 2005, "Reflecting on IMD's forecast model", *Current Science*, 83, 936-938.

Guhathakurta P. and Rajeevan M., 2007, "Trends in the rainfall pattern over India", *Int. J. Climatology,* doi:10.1002/joc.1640.

IITM, 2008, Indian Institute of Tropical Meteorology, Pune, web site http://www.tropmet.res.in.

Jagannathan P., 1960, *Seasonal forecasting in India: A review*, India Meteorological Department, Pune, 80 pp.

Kang I. and Shukla J., 2005, "Dynamical seasonal prediction and predictability of monsoon", in *The Asian Monsoon* [Ed. Wang B.], Springer Praxis Books, 580-612.

Kothawale D. R., Munot A. A. and Borgaonkar H. P., 2007, "Temperature variability over the Indian Ocean and its relationship with Indian summer monsoon rainfall", *Theoretical Applied Climatology*, doi:10.1007/s00704-006-0291-z.

Kripalani R. H. and Kulkarni A., 1997, "Climatic impact of El Nino/La Nina on the Indian monsoon: A new perspective", Weather, 52, 39-46.

Kripalani R. H., Kulkarni A., Sabade S. S. and Khandekar M. L., 2003, "Indian monsoon variability in a global warming scenario", *Natural Hazards*, 29, 189-206.

Krishna Kumar K., Rajagopalan B. and Cane M. A., 1999, "On the weakening relationship between the Indian monsoon and ENSO", *Science*, 284, 2156-2159.

Krishna Kumar K., Rajagopalan B., Hoerling M., Bates G. and Cane M., 2006, "Unravelling the mystery of Indian monsoon failure during El Nino", *Science*, 314, 115-119.

Krishnamurthy V. and Shukla J., 2007, "Intraseasonal and seasonally persisting patterns of Indian monsoon rainfall", *J. Climate*, 20, 3-20.

Krishnamurti T. N. and coauthors, 2000, "Multimodel ensemble forecasts for weather and seasonal climate", *J. Climate*, 13, 4196-4216.

Krishnamurti T. N., 2004, "Weather and seasonal climate prediction of Asian summer monsoon", http://weather.nps.navy.mil, 34 pp.

Krishnamurti T. N., Gnanaseelan C. and Chakraborty A., 2007, "Prediction of the diurnal change using a multimodel superensemble, Part I: Precipitation", *Monthly Weather Review*, 135, 3613-3612.

Kug J., and co-authors, 2005, "Preconditions for El Nino and La Nina onsets and their relation to the Indian Ocean", *Geophysical Research Letters*, 32, doi:10.1029/2004GL021674.

Kulkarni A., Sabade S. S. and Kripalani R. H., 2007, "Association between extreme monsoons and the dipole mode over the Indian subcontinent", *Meteorology Atmopsheric Physics*, 95, 255-268.

MAIRS, 2007, *MAIRS Update*, June 2007 issue, http://www.mairs-essp.org.

Mohanty U. C., Mandal M., Das A. K. and Dimri A. R., 2003, "Mesoscale modeling of convective systems over India: Status and scope", in *Weather and Climate Modelling* [Ed: Singh S. V. et al], New Age International Publishers, New Delhi, 63-76.

Montgomery R. B., 1940, "Reports on the work of G. T. Walker". *Monthly Weather Review*, 39, 1-26.

Mooley D. A., Parthasarathy B., Sontakke N. A. and Munot A. A., 1981, "Annual rain-water over India, its variability and impact on the economy", *J. Climatology*, 1, 167-186.

Normand C. W. B., 1953, "Monsoon seasonal forecasting". *Quarterly J. Royal. Meteorological Society*, 79, 342-346.

Pankaj Kumar, Rupa Kumar K., Rajeevan M. and Sahai A. K., 2007, "On the recent strengthening of the relationship between ENSO and northeast monsoon rainfall over South Asia", *Climate Dynamics*, 28, 49-660.

Parthasarathy B., Sontakke N. A., Munot A. A. and Kothawale D. R., 1987, "Droughts/floods in the summer monsoon rainfall season over different meteorological subdivisions of India for the period 1871-1984", *J. Climatology*, 7, 57-70.

Parthasarathy B., Munot A. A. and Kothawale D. R., 1995, "All India monthly and seasonal rainfall series: 1871-1993". *Theoretical Applied Climatology*, 49, 217-224.

Raj Y. E. A., 2003, "Onset, withdrawal and intra-seasonal variation of northeast monsoon over coastal Tamil Nadu 1901-2000", *Mausam*, 53, 595-614.

Rajeevan M., Pai D. S., Dikshit S. K. and Kelkar R. R., 2004, "IMD's new operational models for long range forecast of southwest monsoon rainfall over India and their verification for 2003", *Current Science*, 86, 422-431.

Rajeevan M, Pai D. S., Anil Kumar R. and Lal B., 2006a, "New statistical models for long-range forecasting of southwest monsoon rainfall over India", *Climate Dynamics*, doi:10.1007/s00382-006-0197-6.

Rajeevan M., 2006, Bhate J., Kale J. D. and Lal B., 2006b, "High resolution daily gridded rainfall data for the Indian region: Analysis of break and active monsoon spells", *Current Science*, 91, 296-306.

Reddy P. R. C. and Salvekar P. S., 2003, "Equatorial Indian Ocean sea surface temperature: A new predictor for seasonal and annual rainfall", *Current Science*, 85, 1600-1604.

Saji N. H., Goswami B. N., Vinayachandran P. N. and Yamagata T., 1999, "A dipole mode in the tropical Indian Ocean", *Nature*, 401, 360-363.

Sanjeeva Rao P. and Sikka D. R., 2005, "Intraseasonal variability of the summer monsoon over the North Indian Ocean as revealed by the BOBMEX and ARMEX field programs", *Pure Applied Geophysics,* 162, 1481-1510.

Sanjeeva Rao P. S. and Sikka D. R., 2007, ""Interactive aspects of the Indian and the African summer monsoon systems", *Pure Applied Geophysics*, 164, 1699-1716.

Slingo J, and Annamalai H., 2000, "1997: The El Niño of the Century and the Response of the Indian Summer Monsoon", *Monthly Weather Review*, 128, 1778-1797.

Solomon S. and coauthors, 2007a, "Patterns (modes) of climate variability", *Technical Summary, Climate Change 2007: The Physical Science Basis. Contribution of Working Group I to the Fourth Assessment Report of the Intergovernmental Panel on Climate Change,* Cambridge University Press, 38-39.

Solomon S. and coauthors, 2007b, "Hierarchy of Global Climate Models", *Technical Summary, Climate Change 2007: The Physical Science Basis. Contribution of Working Group I to the Fourth Assessment Report of the Intergovernmental Panel on Climate Change,* Cambridge University Press, 67.

Solomon S. and coauthors, 2007c, "Evaluation of atmosphere-ocean general circulation models", *Technical Summary, Climate Change 2007: The Physical Science Basis. Contribution of Working Group I to the Fourth Assessment Report of the Intergovernmental Panel on Climate Change,* Cambridge University Press, 59-60.

Walker G. T., 1924, "Correlation in seasonal variations of weather – IX, A further study of world weather", *Indian Meteorological Memoirs*, 24, 275-332. [Reprinted in IMS (1986), 179-240.]

Wang B. and coauthors, 2005, "Fundamental challenge in simulation and prediction of summer monsoon rainfall". *Geophysical Research Letters*, 32, doi:10.1029/2005GL022734.

Webster P. J., Moore A. M., Loschnigg J. P. and Leben R. R., 1999, "Coupled oceanic-atmospheric dynamics in the Indian Ocean during 1997-98", *Nature*, 401, 356-360.

Xavier P. K. and Goswami B. N., 2007a, "A promising alternative to prediction of seasonal mean all India rainfall", *Current Science*, 93, 195-202.

Xavier P. K., Marzin C. and Goswami B. N., 2007b, "An objective definition of the Indian summer monsoon season and a new perspective on the ENSO-monsoon relationship", *Quarterly J. Royal Meteorological Society,* 133, 749-764.

# Subject Index